Atmospheric Aqueous-Phase Chemistry

Atmospheric Aqueous-Phase Chemistry

Editor

Irena Grgić

MDPI • Basel • Beijing • Wuhan • Barcelona • Belgrade • Manchester • Tokyo • Cluj • Tianjin

Editor
Irena Grgić
Department of Analytical Chemistry
National Institute of Chemistry
Ljubljana
Slovenia

Editorial Office
MDPI
St. Alban-Anlage 66
4052 Basel, Switzerland

This is a reprint of articles from the Special Issue published online in the open access journal *Atmosphere* (ISSN 2073-4433) (available at: www.mdpi.com/journal/atmosphere/special_issues/Aqueous_Phase_Chemistry).

For citation purposes, cite each article independently as indicated on the article page online and as indicated below:

LastName, A.A.; LastName, B.B.; LastName, C.C. Article Title. *Journal Name* **Year**, *Volume Number*, Page Range.

ISBN 978-3-0365-7323-6 (Hbk)
ISBN 978-3-0365-7322-9 (PDF)

Contents

About the Editor

Irena Grgić

Irena Grgić obtained her Ph.D. in Chemistry (Atmospheric Chemistry) from the University of Ljubljana, Slovenia, in 1990. Currently, she is a Senior Scientist at the Department of Analytical Chemistry, National Institute of Chemistry, Slovenia. Since 2003, she has also been affiliated with the University of Ljubljana, Faculty of Chemistry and Chemical Technology, from 2015 as a Scientific Councilor.

She has been actively involved in research on atmospheric multiphase processes for 30 years. Her research deals with fundamental processes in atmospheric aqueous-phase chemistry, the physico-chemical characterization and reactivity of aerosols, the formation of secondary organic aerosols, and analytical methods for environmentally important species and involves both field measurements and laboratory experiments.

Preface to "Atmospheric Aqueous-Phase Chemistry"

Liquid water in cloud and fog droplets and in moist aerosol particles is ubiquitous in the atmosphere. Dissolved species from the soluble aerosol fraction, as well as soluble trace gases, undergo chemical reactions in the aqueous phase via different mechanisms, usually yielding different products from those in the gas phase. In addition to their different reactivity, the chemical species solubility determines their fate in the atmosphere, i.e., their involvement in gas-phase or aqueous-phase chemistry.

Numerous studies confirm that the predominant fraction of atmospheric sulfate formed through multiphase oxidation of sulfur (IV) from fossil fuel combustion takes place in cloud droplets, yet there are still unresolved questions concerning sulfate formation under extremely polluted conditions, such as in China.

Recently, it has been recognized that secondary organic aerosol (SOA) mass may also be formed via chemical reactions, in cloud and fog droplets and moist aerosol particles. During atmospheric processing, the primary emitted organic pollutants become more oxidized, less volatile and more water-soluble. Consequently, within the pollutants' lifetime in the atmosphere, aqueous-phase chemistry becomes more and more important for their aging.

This Special Issue of *Atmosphere*, "Atmospheric Aqueous-Phase Chemistry", comprises ten original articles dealing with different aspects of chemistry in atmospheric liquid water. The articles can be divided into three groups: (i) the first group of four articles is mostly based on field measurements and/or combined field and modeling studies giving insights into the chemical characterization of different atmospheric liquid water samples from various environments; (ii) the second set of four articles is focused on studies of aqueous-phase reactivity of some important atmospheric organic compounds; and (iii) the final group comprises two articles based on predictive modeling and/or combined modeling, and laboratory studies providing insight into aqueous secondary organic aerosol (SOA) formation.

I thank all the authors for their valuable contributions, the reviewers for their constructive comments, and the staff of the editorial office at MDPI for their cooperation and support in publishing this issue. I hope that the articles of this Special Issue will stimulate further research to improve the current understanding of atmospheric aqueous-phase chemistry.

Irena Grgić
Editor

atmosphere

Editorial

Atmospheric Aqueous-Phase Chemistry

Irena Grgić

Department of Analytical Chemistry, National Institute of Chemistry, Hajdrihova 19, SI-1000 Ljubljana, Slovenia; irena.grgic@ki.si

Keywords: atmospheric aqueous-phase chemistry; multiphase chemistry; organic pollutants; inorganic species; kinetic studies; mechanistic studies; chemical mechanisms; modeling

Citation: Grgić, I. Atmospheric Aqueous-Phase Chemistry. *Atmosphere* **2021**, *12*, 3. https://dx.doi.org/10.3390/atmos12010003

Received: 16 December 2020
Accepted: 21 December 2020
Published: 23 December 2020

Publisher's Note: MDPI stays neutral with regard to jurisdictional claims in published maps and institutional affiliations.

The Atmosphere Special Issue "Atmospheric Aqueous-Phase Chemistry" comprises ten original articles dealing with different aspects of chemistry in atmospheric liquid water. Liquid water in cloud and fog droplets and in moist aerosol particles is ubiquitous in the atmosphere. Dissolved species from the soluble aerosol fraction as well as soluble trace gases undergo chemical reactions in the aqueous phase via different mechanisms, usually yielding different products from those in the gas phase. In addition to their different reactivity, the chemical species solubility determines their fate in the atmosphere, i.e., their involvement in gas-phase or aqueous-phase chemistry.

The articles of this Special Issue can be divided into three groups: (i) the first is mostly based on field measurements and/or combined field and modeling studies giving insights into the chemical characterization of different atmospheric liquid water samples from various environments [1–4]; (ii) the second group is focused on studies of aqueous-phase reactivity of some important atmospheric organic compounds [5–8]; (iii) and the final group comprises articles based on predictive modeling and/or combined modeling and laboratory studies providing insight into aqueous secondary organic aerosol (SOA) formation [9,10].

The lack of understanding of aerosol liquid water (ALW) and the extent to which this biases $PM_{2.5}$ (particulate matter) risk estimates as related to human exposure to water-soluble pollutants are key knowledge gaps. The associations among ALW, sulfate (as a surrogate for ALW), and changes to $PM_{2.5}$ mass based on filter equilibration were explored across the United States in 2004 by Babila et al. [1]. They found that during sampling and filter equilibration to standard laboratory conditions, ambient ALW mass was removed corresponding to up to 85% of the ambient water mass at most sampling sites. The removal of ALW can induce the evaporation of other semi-volatile compounds present in $PM_{2.5}$ and, thus produce an artifact in PM mass measurements, which is not uniform in space or time [1].

Renard et al. [2] performed statistical analyses (principal component analysis—PCA, agglomerative hierarchical clustering—AHC) of 295 cloud samples collected in the period 2001–2018 at the Puy de Dôme, France (PUY), which resulted in clustering of the cloud samples according to their chemical properties. Further, this classification was combined with a numerical analysis by CAT model (computing advection interpolation of atmospheric parameters and trajectory tool); and, finally, the influence of cloud microphysical properties (liquid water content—LWC and r_e) was investigated by PLS (partial least squares) analysis. The results suggested that cloud chemical composition at PUY was influenced by air mass history including several physicochemical processes (e.g., cloud condensation nuclei (CCN) physical and chemical processes, mass transfer of soluble species, multiphase reactivity), while the influence of cloud microphysics remained minor [2].

Chemical characteristics of organic matter and trace elements (TEs) in rainwater samples collected in an urban area of Zagreb, Croatia were investigated by Orlović-Leko et al. [3]. Their study included the magnitude and variability of TEs, dissolved (DOC)

and particulate organic carbon (POC), and also some selected reactive classes of organic and sulfur species, which can have an important role in the processes of atmospheric and surface waters. Their results showed relatively low concentrations of DOC and TEs; in addition, the daily fluxes of Zn, Pb, Ni, AS and Cd were two to three times lower than those reported for urban sites in Europe.

The presence of reduced sulfur species (RSS) in the water-soluble fraction of fine marine aerosols from the Middle Adriatic area (Rogoznica Lake) was studied by Cvitešić Kušan et al. using voltammetric methods [4]. They confirmed two types of non-volatile RSS species: mercapto-type, which complexes Hg (RS-Hg) and sulfide/S^0-like compounds, which deposit HgS. Their study indicated the presence of mainly mercapto-type compounds in spring samples, which can be potentially associated with biological activity, while both types of compounds were identified in early autumn aerosols.

Intermediate-volatile organic compounds (IVOC) are recognized as important SOA precursors in the gas phase, while aqueous-phase SOA formation from IVOC is still unclear. Liu et al. [5] investigated aqueous oxidation of dibenzothiophene (DBT) as an example of IVOC. They demonstrated that both photo-oxidation by OH radicals and dark aging contribute to aqueous SOAs with yields of 32% and 15%, respectively. In addition, they found that during OH-initiated photochemical reactions of DBT, light-absorbing species are formed, which can significantly contribute to atmospheric brown carbon (BrC).

The study of Vidović et al. [6] demonstrated that aqueous-phase processing of 3-methylcatechol (3MC), as an important precursor to light-absorbing methylnitrocatechols (MNC), in the presence of HNO_2/NO_2^-, both under sunlight and in the dark, may significantly contribute to secondary BrC formation. Under illumination, the degradation of 3MC is faster than in the dark, but the yield of the main two products of the dark reaction (3-methyl-5-nitrocatechol, 3M5NC and 3-metyl-4-nitrocatechol, 3M4NC) is low, suggesting different degradation pathways of 3MC in the sunlight. Besides the primary reaction products (3M5NC and 3M4NC), second-generation products with absorption above 400 nm were also confirmed. The characteristic mass absorption coefficient (MAC) values were found to increase with increasing NO_2^-/3MC concentration ratio and decrease with increasing wavelength.

In their work, Rudziński and Szmigielski [7] elucidated how quickly different nitrophenols, as common and hazardous environmental pollutants, are removed from the atmosphere via aqueous-phase reactions with sulfate radical anions (SO_4^-). Their rough estimations showed that the gas-phase reactions of 2-nitrophenol (2-NP) with OH or NO_3 radicals dominate over the aqueous-phase reaction with SO_4^- in deliquescent aerosol and haze. In cloud, rain, and fog waters, aqueous-phase reactions of 2-NP with SO_4^- radicals dominate if their concentrations are similar to those of OH or NO_3 radicals.

In the work of Schaefer et al. [8], the pH- and temperature-dependent oxidation reactions of dicarboxylic acids (succinic and pimelic acid), as important trace constituents in the troposphere present mostly in the particle phase, were investigated to determine the OH radical rate constants. They observed that the rate constant increases with increasing pH due to the deprotonation of the carboxylic group ($k(AH_2) < k(AH^-) < k(AH^{2-})$). The atmospheric lifetime of dicarboxylic acids was estimated to range from a few minutes under aerosol conditions to several hours under cloud conditions. The oxidation reaction by OH can act as a sink for dicarboxylic acids as well as a source of smaller dicarboxylic acids or multifunctional compounds.

The contribution of dark SOA formation from carbonyl-containing compounds (CVOCs) to overall aqueous aerosol optical properties is expected to be significant. The multiple pathways taking place in such systems and different chromophoricity of individual products make it complicated to reliably model the kinetics. In their work, Fan et al. [9] proposed an alternative method of representing UV–Vis absorbance spectra as a composite of Gaussian lineshape functions to infer kinetic information. Their data and model parameters were focused on the light-absorbing properties of glyoxal, methylglyoxal, and glycolaldehyde in ammonium sulfate environments, selected due to their notable contributions to aqueous SOA mass.

Although it is recognized that SOAs are formed during chemical processing of organic material in aqueous aerosols and cloud water, it is not yet clear how much each of these processes contributes to total aerosol mass. In their work, Tsui et al. [10] used a photochemical box model (Gas-Aerosol Model for Mechanism Analysis—GAMMA 5.0) with coupled gas- and aqueous-phase chemistry to investigate the impact of aqueous reactions on isoprene epoxydiol (IEPOX) SOA formation in both aerosol and clouds, including cycling between the two phases upon the evaporation or condensation of water. Their results indicate that aqueous-phase chemistry in clouds (pH 4 or lower) may be a more substantial source of IEPOX SOA than formerly believed. This is primarily attributed to the relatively higher uptake of IEPOX to clouds compared with aerosol. Formation in clouds could account for the discrepancies between predicted IEPOX SOA mass from atmospheric models and measured ambient IEPOX SOA mass or observations of IEPOX SOA in locations where mass transfer limitations are expected in aerosols.

Funding: This research received no external funding.

Acknowledgments: I would like to thank all the authors for their valuable contributions, the reviewers for their constructive comments, and the staff of the editorial office MDPI for good cooperation and support in publishing this issue. I hope that articles of this Special Issue will stimulate further research to improve the current understanding of atmospheric aqueous-phase chemistry.

Conflicts of Interest: The authors declare no conflict of interest.

References

1. Babila, J.E.; Carlton, A.G.; Hennigan, C.J.; Ghate, V.P. On Aerosol Liquid Water and Sulfate Associations: The Potential for Fine Particulate Matter Biases. *Atmosphere* **2020**, *11*, 194. [CrossRef]
2. Renard, P.; Bianco, A.; Baray, J.-L.; Bridoux, M.; Delort, A.-M.; Deguillaume, L. Classification of Clouds Sampled at the Puy de Dôme (France) Based on Chemical Measurements and Air Mass History Matrices. *Atmosphere* **2020**, *11*, 732. [CrossRef]
3. Orlović-Leko, P.; Vidović, K.; Ciglenečki, I.; Omanović, D.; Dutour Sikirić, M.; Šimunić, I. Physico-Chemical Characterization of an Urban Rainwater (Zagreb, Croatia). *Atmosphere* **2020**, *11*, 144. [CrossRef]
4. Cvitešić Kušan, A.; Frka, S.; Ciglenečki, I. Electrchemical Evidence of non-Volatile Reduced Sulfur Species in Water-Soluble Fraction of Fine Marine Aerosols. *Atmosphere* **2019**, *10*, 674. [CrossRef]
5. Liu, Y.; Lu, J.; Chen, Y.; Liu, Y.; Ye, Z.; Ge, X. Aqueous-Phase Production of Secondary Organic Aerosols from Oxidation of Dibenzothiophene (DBT). *Atmosphere* **2020**, *11*, 151. [CrossRef]
6. Vidović, K.; Kroflič, A.; Šala, M.; Grgić, I. Aqueous-Phase Brown Carbon Formation from Aromatic Precursors under Sunlight Conditions. *Atmosphere* **2020**, *11*, 131. [CrossRef]
7. Rudziński, K.J.; Szmigielski, R. Aqueous Reactions of Sulfate Radical-Anions with Nitrophenols in Atmospheric Context. *Atmosphere* **2019**, *10*, 795. [CrossRef]
8. Schaefer, T.; Wen, L.; Estelmann, A.; Maak, J.; Herrmann, H. pH- and Temperature-Dependent Kinetics of the Oxidation Reactions of OH with Succinic and Pimelic Acid in Aqueous Solution. *Atmosphere* **2020**, *11*, 320. [CrossRef]
9. Fan, M.; Ma, S.; Ferdousi, N.; Dai, Z.; Woo, J.L. Modeling of Carbonyl/Ammonium Sulfate Aqueous Brown Carbon Chemistry via UV/Vis Spectral Decomposition. *Atmosphere* **2020**, *11*, 358. [CrossRef]
10. Tsui, W.G.; Woo, J.L.; McNeill, V.F. Impact of Aerosol-Cloud Cycling on Aqueous Secondary Organic Aerosol Formation. *Atmosphere* **2019**, *10*, 666. [CrossRef]

atmosphere

MDPI

Article

On Aerosol Liquid Water and Sulfate Associations: The Potential for Fine Particulate Matter Biases

Jonathon E. Babila [1], Annmarie G. Carlton [1,*], Christopher J. Hennigan [2] and Virendra P. Ghate [3]

1 Department of Chemistry, University of California, Irvine, CA 92617, USA; jbabila@uci.edu
2 Department of Chemical, Biochemical, and Environmental Engineering, University of Maryland, Baltimore County, 1000 Hilltop Circle, Baltimore, MD 21250, USA; hennigan@umbc.edu
3 Environmental Science Division, Argonne National Laboratory, Argonne, IL 60439, USA; vghate@anl.gov
* Correspondence: agcarlto@uci.edu

Received: 16 January 2020; Accepted: 5 February 2020; Published: 12 February 2020

Abstract: In humid locations of the Eastern U.S., sulfate is a surrogate for aerosol liquid water (ALW), a poorly measured particle constituent. Regional and seasonal variation in ALW–sulfate relationships offers a potential explanation to reconcile epidemiology and toxicology studies regarding particulate sulfur and health endpoints. ALW facilitates transfer of polar species from the gas phase to the particle phase and affects particle pH and metal oxidation state. Though abundant and a potential indicator of adverse health endpoints, ALW is largely removed in most particulate matter measurement techniques, including in routine particulate matter ($PM_{2.5}$) networks that use federal reference method (FRM) monitors, which are used in epidemiology studies. We find that in 2004, a typical year in the available record, ambient ALW mass is removed during sampling and filter equilibration to standard laboratory conditions at most (94%) sites, up to 85% of the ambient water mass. The removal of ALW can induce the evaporation of other semi-volatile compounds present in $PM_{2.5}$, such as ammonium nitrate and numerous organics. This produces an artifact in the PM mass measurements that is, importantly, not uniform in space or time. This suggests that $PM_{2.5}$ epidemiology studies that exclude ALW are biased. This work provides a plausible explanation to resolve multi-decade discrepancies regarding ambient sulfate and health impacts in some epidemiological and toxicological studies.

Keywords: $PM_{2.5}$; air quality; aerosol liquid water

---

1. Introduction

More than one million life years are lost in the U.S. due to fine particulate matter ($PM_{2.5}$) exposure during a single year. [1] Studies link $PM_{2.5}$ mass with a variety of adverse health consequences [2]. Sulfate, a major $PM_{2.5}$ constituent, is often positively correlated with increased risk in many epidemiology studies [3,4], including the Harvard six cities study [5], but not all [6]. Toxicology studies do not reveal mechanistic evidence for a direct association between sulfate and health endpoints at atmospherically relevant concentrations [6,7]. Inconsistent epidemiological findings and lack of a definitive toxicological explanation are a conundrum.

A plausible explanation for the discrepancies is that sulfate is not toxic by itself, but a surrogate for co-produced constituent(s) not well characterized in $PM_{2.5}$ measurements [6,8]. Historically, many epidemiological studies relating $PM_{2.5}$ to health endpoints rely on data from fixed location 'central' monitors (e.g., [9]). Sulfate is not volatile and exhibits less spatial variability than other $PM_{2.5}$ constituents [10]. Correlations between indoor (where most people spend most time) and outdoor sulfate are high [11], and measurement errors and impact on risk ratios [12] are lower for sulfate than for total $PM_{2.5}$ mass. Therefore, such central monitor approaches should work well for sulfate.

However, in a review of time series studies investigating the impacts of $PM_{2.5}$ and sulfate on various health endpoints, Reiss et al. (2008) (using re-estimated relative risk ratios to account for misleading results from S-Plus' default convergence criteria in the Generalized Additive Model), found little to no support for a causal association of sulfate and increased health risk at ambient concentrations. In their analysis of reported mass concentrations at central monitor locations in routine air quality networks for 2004, Reiss et al. note that, for cities across the U.S., statistical correlations between sulfate and $PM_{2.5}$ vary, reducing the power of epidemiologic studies to elucidate association between specific $PM_{2.5}$ constituents and health risk. Such arguments could be used to advocate for the loosening of successful sulfur-related regulations. However, there is a plausible mechanism by which sulfate may impact human health through indirect ways not assessed in many current approaches [8,13], namely by increasing the amount of aerosol liquid water (ALW) and facilitating transfer to the particle phase of polar, water-soluble organic species that are potentially toxic and induce harm.

Sulfate is a hygroscopic species that promotes water uptake. Although ALW is an abundant and ubiquitous $PM_{2.5}$ chemical constituent [14], it is removed during $PM_{2.5}$ mass sampling at routine air quality network sites [15], and is vaporized, and not detected by state-of-the-art particle measurement technologies such as aerosol mass spectrometers [16]. ALW facilitates transfer from the gas phase to the particle phase of polar organic species [17], which are highly soluble in a polar solvent, such as water. ALW alters the amount and chemical composition of particulate organic mass [18], potentially in ways important for human health [19]. For example, secondary organic aerosol (SOA) derived from isoprene epoxydiols is only formed when ALW is present [20]. This SOA induces expression of oxidative stress response genes in human bronchial epithelial cells, which is not seen in control experiments with sulfate aerosol [21–24]. Further, in humid locations ALW correlates with water-soluble toxic species (e.g., Fe, Cu) [25], reactive oxygen species (ROS) [25], particle oxidative potential [8], and sulfate [26]. Sulfate is the most widely reported species routinely measured and reported by air quality networks. Positive sulfate–health associations are noted in the Harvard six cities study (i.e., all humid Eastern U.S. locations, Figure 1), and generally observed in areas of the U.S. where ALW is predicted to be high [17], (e.g., Detroit [27], and New Jersey [28]) but not in areas where ALW is expected to be low [17], (e.g., Santa Clara [29], and Phoenix [30]). Associations between ALW and health end points are difficult to quantitatively assess with many current approaches.

Central monitors in air quality networks across the U.S. often measure $PM_{2.5}$ surface mass concentrations according to the federal reference method (FRM), a filter-based mass measurement collected and averaged over 24-h (midnight to midnight). Quality assurance criteria require that after collection, FRM $PM_{2.5}$ filters are equilibrated for a minimum of 24-h at 20–23 °C (±1 °C) and 30–40% (±5%) relative humidity (RH) [31]. Hourly $PM_{2.5}$ mass concentration measurements using federal equivalent methods (FEM) also evaporate ALW from ambient particles prior to collection [15]. This protocol perturbs collected particles from ambient temperature and RH to standard laboratory conditions, and it is well documented that sampling artifacts arise due to the volatilization of organic material during filter sampling [32,33]. FRM sampling and equilibration procedures, in particular for water-soluble organic species, are also prone to this bias [34]. Ambient SOA equilibration experiments to laboratory conditions (RH = 35%) by El-Sayad et al. (2016) for samples collected in Baltimore, MD (i.e., humid, Eastern U.S.) find water-soluble organic matter evaporation losses are in excess of 5 $\mu g\ m^{-3}$ in some cases. This is consistent with other particulate organic carbon findings by Zhang et al., (2012) for samples collected in and near Atlanta, GA (Eastern U.S.) [35]. Loss of water-soluble organic species may obscure interpretation of the health impacts of inhaled ambient $PM_{2.5}$. Organic species are documented to exhibit redox activity and catalyze ROS generation upon deposition in lung fluid to induce adverse health responses, including oxidative stress, cell death, and disease [36]. Standard equilibration procedures in the FRM ensure consistency among reported mass concentrations used in $PM_{2.5}$ National Ambient Air Quality Standard (NAAQS) attainment decisions for the States. For example, humid locations will have more ALW mass than arid regions for the same amount of dry $PM_{2.5}$ of a given chemical composition. ALW mass concentrations per unit of measured and reported $PM_{2.5}$ 'dry' mass are highest in the Eastern U.S. [37]. Equilibration to standard conditions normalizes the ALW content in reported $PM_{2.5}$ mass across the U.S. for a strict set of laboratory conditions, asserting fairness to the mass-based NAAQS attainment process. However, data from FRM measurements are used in epidemiology studies [38], suggesting that the measurement bias associated with ALW evaporation may influence the analysis of statistical links to health endpoints.

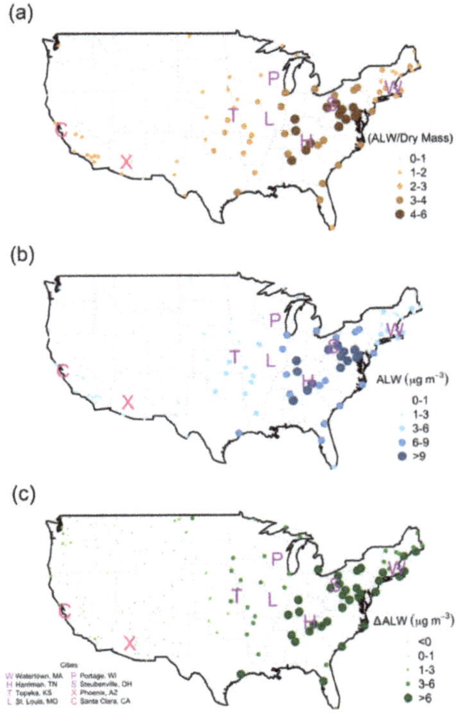

Figure 1. Yearly average in 2004 at Interagency Monitoring of PROtected Visual Environments (IMPROVE) monitoring sites of (**a**) ratio of aerosol liquid water (ALW) mass per measured $PM_{2.5}$ 'dry' mass (**b**) ambient ALW mass concentrations and (**c**) equivalent concentration loss upon filter equilibration to standard laboratory conditions. Point color and diameter are relative to ALW per dry mass ratio (orange), ALW mass (blue), and ALW mass difference (green, red). Green indicates ALW loss during equilibration and red indicates ALW gain.

This work explores the relationships among sulfate, ALW, and changes to $PM_{2.5}$ mass upon filter equilibration across the U.S. with a focus on 2004, the year in the sulfate-epidemiology review by Reiss [6]. We characterize ALW and changes to $PM_{2.5}$ mass and dry versus liquid fractionation through estimates at ambient and laboratory conditions. Precise chemical characterization of compounds associated with or formed in ALW, and then lost during filter sampling and/or equilibration remains a critical open question. The lack of understanding in ALW and the extent to which this biases $PM_{2.5}$ risk estimates as they relate human exposure to water-soluble air pollution are key knowledge gaps.

2. Methods

We estimate ALW across the contiguous U.S. using surface mass concentrations of SO_4^{2-} and NO_3^- at $PM_{2.5}$ monitoring locations in the Interagency Monitoring of PROtected Visual Environments (IMPROVE) network [39]. The estimation technique has been described in detail elsewhere [40]. Briefly,

IMPROVE collects 24-h PM filter measurements every three days. Teflon filters are analyzed for gravimetric fine mass, elemental concentration, and light absorption, while nylon filters are analyzed for the anions sulfate, nitrate, and chloride using ion chromatography. Data was downloaded from the IMPROVE public archive [41] on 13 July 2015 for 203 unique locations (Figure S1) from 1988 to 2014, and the database includes additional measurement details. There is some variability in the number and location of IMRPOVE sites year to year. In 2004, 158 IMPROVE sites with a minimum of 2 reported measurements per month are used for this analysis. Meteorology measurements are not available at IMPROVE sites, therefore, we employ the North American Regional Reanalysis (NARR) model [42], paired in space and time with monitoring sites, to obtain relative humidity (RH) and temperature data. We calculate ALW using NARR meteorology, IMRPOVE particle chemical composition, and the thermodynamic equilibrium model ISORROPIAv2.1 [43] that describes gas-to-particle partitioning of inorganic species, including water (Figure S2). We apply ISORROPIA as a box model where IMPROVE chemically characterized $PM_{2.5}$ and NARR meteorology are inputs. We assume metastable particles, and that measured sulfate is in the form of ammonium sulfate, and nitrate is in the form of ammonium nitrate. Ammonia (NH_3), dust, and organic compounds are not considered in the ALW estimates due to inconsistencies in measurements among the sites, an inability to accurately predict concentrations, and/or large uncertainties in intrinsic properties such as water uptake relationships. The IMPROVE monitoring network routinely measures ammonium (NH_4^+) mass, but not gas-phase ammonia. Gas-phase predictions of ammonia by atmospheric models are not accurate [44], and therefore inappropriate to use as input to the thermodynamic equilibrium is pH-dependent for which accurate constraints are elusive in routine monitoring data [45] Neglect of ammonia in our ALW mass estimates likely underestimates ALW content. Dust contains crustal species (e.g., Ca^{2+}, K^+, Mg^{2+}) which can form insoluble species (e.g., $CaSO_4$). Neglect of dust likely overestimates ALW mass. Organic species vary greatly in their effects on ALW [46,47], and detailed chemical characterization of the organic fraction of $PM_{2.5}$ is not available. Neglect of these species introduces uncertainty, and while absolute mass concentrations are uncertain, we are confident that overall trends are qualitatively robust in space and time [37,40] Laboratory conditions for estimated ALW are 21.5 °C and 35% RH for all sites on all measurement days. These are the central values established for equilibration of FRM $PM_{2.5}$ filters, and we calculate laboratory ALW estimates for the central values, as an idealized case [31].

3. Results

In 2004, the ratio of ALW mass per measured $PM_{2.5}$ 'dry' mass, average ALW mass concentrations, and water loss upon filter equilibration to standard laboratory conditions are greatest in the Eastern U.S. (Figure 1), where both sulfate mass and RH are relatively high (Figure S3). This year is highlighted in a review of sulfate–health endpoint associations and represents typical conditions that may partly explain discrepancies in the literature among epidemiological and toxicological studies. The pattern observed for 2004 persists throughout the entire IMPROVE record analyzed in this work (Video S1, Video S2), though sulfate concentrations and absolute values of ALW mass concentrations are decreasing over past decades, consistent with findings in the Southeast U.S. [40]. The patterns in ALW and ALW:$PM_{2.5}$ ratios estimated here are independently consistent with remote optical measurements that demonstrate higher RH-dependent extinction per unit of $PM_{2.5}$ dry mass in the Eastern U.S. [37]. Note the spatial similarity of the humid Eastern U.S. where these ALW patterns are observed with epidemiology studies that find statistically discernible relationships among sulfate mass concentrations and health endpoints. The locations of all the Harvard six cities are in areas with relatively high ALW mass concentrations, as are other cities such as Detroit [27], New Jersey [28], and Toronto [48]. Locations where statistically robust associations among sulfate and health endpoints are not found, for example in Santa Clara, California and Phoenix, Arizona, are arid and estimated ALW is low (Figure 1). It is noted that nitrate is more hygroscopic than sulfate, but nitrate mass concentrations are generally small (<1 μg m^{-3} in most locations) and where they are highest, it is in the areas of the U.S. where RH is low (Figure S3). Nitrate impacts on ALW are therefore generally less than for sulfate.

While organic material is lost during FRM sampling and analysis approaches [32,33], loss of water mass is greater. ALW mass is lost during equilibration from ambient to laboratory conditions at nearly all monitoring locations (94%), with the exception of nine sites in the Southwestern U.S. where RH is low (Figure 1b, Figure S3). At all Eastern U.S. sites, throughout the entire analyzed time period, ALW mass is abundant (Figure S4, Video S1), and there is a positive difference between estimated ALW for ambient and lab conditions, (Video S2) indicating systematic ALW evaporation from filters occurs broadly for this region. Average ALW mass concentrations at most IMPROVE monitoring locations in the Eastern U.S. are >9 $\mu g\ m^{-3}$ and lose the equivalent ALW mass concentration of more than 6 $\mu g\ m^{-3}$ upon filter equilibration (Figure 1). This finding suggests that on average for the Eastern U.S. in 2004, ALW contributed 3 $\mu g\ m^{-3}$ to reported $PM_{2.5}$ 'dry' mass. The site with the largest ambient ALW mass concentrations and greatest amount of water loss in 2004 is Martha's Vineyard in Massachusetts (Table 1, Figure 2). Average ALW loss at Martha's Vineyard in 2004 was 17.5 $\mu g\ m^{-3}$ or 85% of ambient water mass. ALW mass concentrations in the Eastern U.S. are substantially more abundant than the Western U.S. The majority of western sites exhibit annual average ALW mass concentrations < 3 $\mu g\ m^{-3}$ in 2004. The site with the least amount of water that year is Lava Beds National Monument in Northern California, 0.6 $\mu g\ m^{-3}$ (Table 1, Figure 2). The site with the largest increase in ALW mass during filter equilibration in 2004 is Death Valley National Park in Central California, where it is estimated that 0.2 $\mu g\ m^{-3}$ of ALW was gained during the laboratory filter equilibration (Table 1, Figure 2). The average ambient RH at Death Valley was 28%, which is less than the 35% used for ALW estimates at laboratory conditions, explaining the estimated ALW enhancement.

Table 1. The 2004 annual average at IMPROVE locations discussed here. Ambient and lab particles are idealized representations to indicate change due to filter equilibrations. Pie chart size is proportional by mass between ambient and lab conditions for an individual location.

City	Sulfate ($\mu g\ m^{-3}$)	Nitrate ($\mu g\ m^{-3}$)	Ambient Water ($\mu g\ m^{-3}$)	Lab Water ($\mu g\ m^{-3}$)	Relative Humidity (%)	Ambient Particle and Mass ($\mu g\ m^{-3}$)	Lab Particle and Mass ($\mu g\ m^{-3}$)
Martha's Vineyard, MA	2.57	0.55	17.5	2.57	85.8	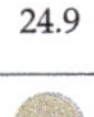 21.6	6.66
Brigantine, NJ	3.72	0.87	11.7	3.71	79.7	 17.9	 9.90
Washington D.C.	4.84	1.57	16.0	4.82	79.7	 24.9	 13.7
Detroit, MI	3.42	2.60	9.78	3.40	76.5	 18.4	 12.1
Phoenix, AZ	1.02	1.05	0.95	1.02	31.1	 5.92	 5.99
Death Valley, CA	0.84	0.23	0.66	0.84	28.1	 2.28	 2.46
Lava Beds, CA	0.35	0.12	0.56	0.35	60.7	 2.01	 1.89

Note: Pie chart colors denote mass fraction of dry material (brown) and water (blue).

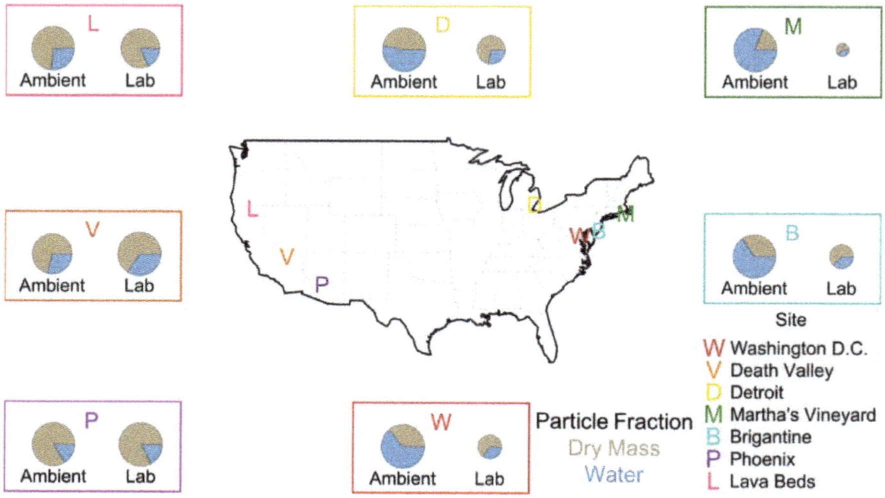

Figure 2. Differences between ambient and laboratory-equilibrated $PM_{2.5}$ absolute mass and constituent fractional contributions. Idealized particles are drawn to scale by mass for ambient and laboratory conditions for each city, but not between cities.

Fractionation of total particle mass that is 'dry' versus water changes dramatically upon filter equilibration in most locations. For all IMPROVE sites in the Eastern U.S. highlighted in Figure 2, Detroit, Martha's Vineyard, Brigantine, and Washington D.C., ambient particulate matter mass is predominantly (>50%) liquid water. After equilibration, the dry fraction dominates total mass. Notable exceptions occur in arid regions of the Western U.S. with low RH, such as Death Valley, and Phoenix, AZ (Figure 2) which experienced ALW gain from ambient to filter equilibration conditions. It should be noted that, in part, the intention of standardized laboratory conditions for equilibration of $PM_{2.5}$ filter samples is to remove liquid water from reported mass values used in attainment considerations. In that respect, standardized mass concentrations normalized to prescribed temperature and RH, works as intended.

In locations where RH and sulfate mass concentrations are abundant, ambient ALW mass concentrations are highest and exhibit the largest temporal variability (Figure 3, Videos S3 and S4). In 2004 in Washington D.C. median ALW mass concentrations vary from < 10 $\mu g\ m^{-3}$ in the winter months to >25 $\mu g\ m^{-3}$ in summer, consistent with decadal seasonal ALW patterns in this region [49] Annual average RH for the region was nearly 80% year round and exhibited little temporal variability relative to sulfate mass. Sulfate mass concentrations increase by a factor of four from summer to winter. Summertime maxima in sulfate mass concentrations in the Eastern U.S. are driven largely by changing SO_2 emissions related to electricity demand, which peaks in summer on High Electricity Demand Days (HEDD) [50,51]. Temporal variance (σ = 7.4 $\mu g\ m^{-3}$) in ALW for Washington D.C. in 2004 is relatively large and driven primarily by the strong seasonal trend of sulfate (Figure 3) and not RH. Sufficiently high year-round RH suggests water vapor mixing ratios do not limit ALW formation in that location and that sulfate has the determining effect on ALW mass concentrations. In contrast, in 2004 for Phoenix, AZ, RH and sulfate mass concentrations are lower (e.g., <50% RH and <2 $\mu g\ m^{-3}$). Temporal variance over the year for ALW mass is small (σ = 0.4 $\mu g\ m^{-3}$), in part, because sulfate mass concentrations and RH are out of phase. In Phoenix, RH is greatest in winter when ambient temperatures are low, and sulfate mass concentrations are highest in summer, same as for D.C. This suggests ambient water mixing ratios can sometimes be limiting and exert a determining impact on ALW mass concentrations, in sharp contrast to humid locations such as D.C. where RH is sufficiently high year round to facilitate ALW formation. Note that statistically robust associations between sulfate and health endpoints have been found in Eastern U.S. locations (e.g., Detroit [27], New Jersey [28]) and not in Phoenix, AZ, where sulfate mass concentrations are also substantially lower [6].

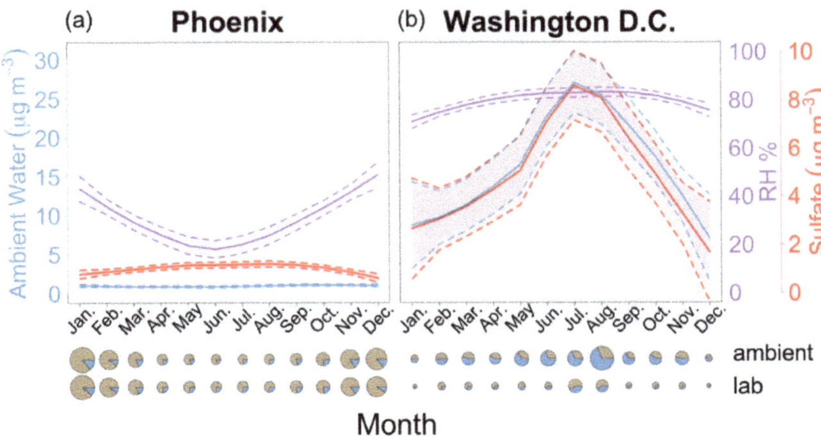

Figure 3. Mass concentrations of aerosol liquid water and particulate sulfate are abundant and positively associated in Washington D.C., and not Phoenix, AZ in 2004. 2004 monthly mean values of ambient water content, relative humidity, and sulfate (**a**) (left) Phoenix, Arizona, and (**b**) (right) Washington D.C. Each variable is presented as the mean (solid line) and 95% confidence interval (shaded area and dotted lines). Pie charts illustrate the monthly average dry mass (brown) and water fractions (blue) for each month in each city at ambient and lab conditions. Pie chart diameter is relative to the largest monthly average mass concentration of each city.

4. Discussion

FRM-collected $PM_{2.5}$ mass is perturbed from ambient conditions upon filter equilibration. Reported mass concentrations are biased low for most locations, including the 25 largest metropolitan areas in the United States (Table S1). Further, collected PM composition is chemically different from the ambient particulate matter to which individuals are exposed, potentially imparting bias in estimated risk ratios calculated in epidemiology studies that use $PM_{2.5}$ FRM measurements. Discrepancies in epidemiology studies linking sulfate to health endpoints may be partly explained by mass and chemical differences in ambient vs. laboratory $PM_{2.5}$ that are induced by evaporation of ALW and organic material. Further, in the absence of plausible toxicological mechanistic explanations for sulfate-induced injury it is important to note that adverse health endpoints associated with ambient sulfate may arise from other particle characteristics not explored here, such as pH or toxic metal oxidation state [8]. In the IMPROVE data set analyzed here, sulfate is correlated with water-soluble transition metals Al, Fe, Pb and Mn ($R^2 > 0.5$) at the IMPROVE site in Lava Beds, CA for 2004 (Table S2). However, for cities in this general location (Santa Clara and Phoenix), where sulfate concentrations are less than 2 $\mu g\ m^{-3}$ year round, statistically robust sulfate–health relationships are not observed in epidemiological studies [29,30]. Another complication is that partitioning of semi-volatile organic material to ALW exhibits a seasonal trend in humid areas [52], indicating that loss of organic material during equilibration is not uniform or easily parameterized from ALW data. Particulate matter in arid Southwestern U.S. locations also contains WSOC that varies by season and is associated with enhanced moisture, temperature, and ozone that may work synergistically to cause health impacts [53].

The absolute and relative amount of ALW loss and bias in FRM-reported $PM_{2.5}$ changes spatially across the U.S., and for a given location throughout the year (Figure 3). There is not a single bias correction that works for all times and locations. Further, decadal decreases in sulfate and ALW mass concentrations in humid and dry locations, alike, are documented in the literature and traceable to environmental regulations, increasing efficiencies and economic factors (Figures S5 and S6) [54]. Observational evidence in visibility measurements is supportive of sulfate-driven ALW decadal trends [55], in particular in the absence of similar RH trends, suggesting that repeating epidemiology studies in certain locations may change interpretation of sulfate–health endpoint relationships if ALW is included in the analysis. These findings qualitatively translate more broadly to other locations outside of the U.S.

Supplementary Materials: The following are available online at http://www.mdpi.com/2073-4433/11/2/194/s1, Figure S1: IMPROVE network monitoring locations, Figure S2: Flow chart summarizing the method for calculating aerosol liquid water, Figure S3: 2004 annual average of sulfate mass, nitrate mass, and RH% at all IMMPROVE network monitoring locations, Video S1: GIF of 1988 to 2013 yearly average ALW for IMPROVE monitoring sites, Video S2: GIF of 1988 to 2013 yearly average difference between ambient and lab ALW for IMPROVE monitoring sites, Figure S4: Defined Eastern and Western U.S., Video S3: GIF of 2004 ALW monthly average for IMPROVE monitoring sites, Video S4: GIF of 2004 monthly average difference between ambient and lab ALW for IMPROVE monitoring sites, Table S1: Chemical speciation data and calculations of aerosol liquid water content at ambient and lab conditions for the 25 largest U.S. cities discussed in Reiss et al., Table S2: 2004 annual average correlations of particulate sulfate and transition metals for a subset of cities, Figure S5: U.S. average ambient aerosol liquid water for 2001 to 2013, Figure S6: Yearly average comparing ambient water content to relative humidity, and ambient water to sulfate mass for Phoenix, AZ, and Washington D.C.

Author Contributions: Data curation, V.P.G.; Formal analysis, J.E.B.; Methodology, V.P.G.; Supervision, A.G.C.; Visualization, J.E.B.; Writing—original draft, A.G.C.; Writing—review & editing, A.G.C. and C.J.H. All authors have read and agreed to the published version of the manuscript.

Funding: This research was funded, in part by NSF grant numbers #1719252 and #1719245, NASA grant number #80NSSC19K0987.

Acknowledgments: Publicly available data from IMPROVE and NARR were necessary for the ALW calculations. Khoi Nguyen and Amy Christiansen contributed to build the ALW dataset.

Conflicts of Interest: The authors declare no conflicts.

References

1. Fann, N.; Lamson, A.D.; Anenberg, S.C.; Wesson, K.; Risley, D.; Hubbell, B.J. Estimating the National Public Health Burden Associated with Exposure to Ambient $PM_{2.5}$ and Ozone. *Risk Anal.* **2012**, *32*, 81–95. [CrossRef] [PubMed]
2. Burnett, R.; Chen, H.; Szyszkowicz, M.; Fann, N.; Hubbell, B.; Pope, C.A.; Apte, J.S.; Brauer, M.; Cohen, A.; Weichenthal, S.; et al. Global estimates of mortality associated with long-term exposure to outdoor fine particulate matter. *Proc. Natl. Acad. Sci. USA* **2018**, *115*, 9592–9597. [CrossRef] [PubMed]
3. Dockery, D.W.; Cunningham, J.; Damokosh, A.I.; Neas, L.M.; Spengler, J.D.; Koutrakis, P.; Ware, J.H.; Raizenne, M.; Speizer, F.E. Health effects of acid aerosols on North American children: Respiratory symptoms. *Environ. Health Perspect.* **1996**, *104*, 500–505. [CrossRef] [PubMed]
4. Raizenne, M.; Neas, L.M.; Damokosh, A.I.; Dockery, D.W.; Spengler, J.D.; Koutrakis, P.; Ware, J.H.; Speizer, F.E. Health effects of acid aerosols on North American children: Pulmonary function. *Environ. Health Perspect.* **1996**, *104*, 506–514. [CrossRef]
5. Dockery, D.W.; Pope, C.A.; Xu, X.; Spengler, J.D.; Ware, J.H.; Fay, M.E.; Ferris, B.G.; Speizer, F.E.N. An Association between Air Pollution and Mortality in Six U.S. Cities. *N. Engl. J. Med.* **1993**, *329*, 1753–1759. [CrossRef]
6. Reiss, R.; Anderson, E.L.; Cross, C.E.; Hidy, G.; Hoel, D.; McClellan, R.; Moolgavkar, S. Evidence of health impacts of sulfate- and nitrate-containing particles in ambient air. *Inhal. Toxicol.* **2007**, *19*, 419–449. [CrossRef]
7. Schlesinger, R.B. The health impact of common inorganic components of fine particulate matter ($PM_{2.5}$) in ambient air: A critical review. *Inhal. Toxicol.* **2007**, *19*, 811–832. [CrossRef]
8. Fang, T.; Guo, H.Y.; Zeng, L.H.; Verma, V.; Nenes, A.; Weber, R.J. Highly Acidic Ambient Particles, Soluble Metals, and Oxidative Potential: A Link between Sulfate and Aerosol Toxicity. *Environ. Sci. Technol.* **2017**, *51*, 2611–2620. [CrossRef]
9. Laden, F.; Schwartz, J.; Speizer, F.E.; Dockery, D.W. Comments on the updated Harvard Six Cities study. *Am. J. Respir. Crit. Care Med.* **2006**, *174*, 722–724. [CrossRef]

10. Baker, K.; Scheff, P. Assessing meteorological variable and process relationships to modeled $PM_{2.5}$ ammonium nitrate and ammonium sulfate in the central United States. *J. Appl. Meteorol. Climatol.* **2008**, *47*, 2395–2404. [CrossRef]
11. Landis, M.S.; Norris, G.A.; Williams, R.W.; Weinstein, J.P. Personal exposures to $PM_{2.5}$ mass and trace elements in Baltimore, MD, USA. *Atmos. Environ.* **2001**, *35*, 6511–6524. [CrossRef]
12. Goldman, G.T.; Mulholland, J.A.; Russell, A.G.; Srivastava, A.; Strickland, M.J.; Klein, M.; Waller, L.A.; Tolbert, P.E.; Edgerton, E.S. Ambient Air Pollutant Measurement Error: Characterization and Impacts in a Time-Series Epidemiologic Study in Atlanta. *Environ. Sci. Technol.* **2010**, *44*, 7692–7698. [CrossRef] [PubMed]
13. Carlton, A.G.; Pye, H.O.T.; Baker, K.R.; Hennigan, C.J. Additional benefits of federal air-quality rules: Model estiamtes of controllable biogenic secondary organic aerosol. *Environ. Sci. Technol.* **2018**, *52*, 9254–9265. [CrossRef] [PubMed]
14. Nguyen, T.K.V.; Zhang, Q.; Jimenez, J.L.; Pike, M.; Carlton, A.G. Liquid Water: Ubiquitous Contributor to Aerosol Mass. *Environ. Sci. Technol. Lett.* **2016**, *3*, 257–263. [CrossRef]
15. Noble, C.A.; Vanderpool, R.W.; Peters, T.M.; McElroy, F.F.; Gemmill, D.B.; Wiener, R.W. Federal reference and equivalent methods for measuring fine particulate matter. *Aerosol Sci. Technol.* **2001**, *34*, 457–464. [CrossRef]
16. Jimenez, J.L.; Jayne, J.T.; Shi, Q.; Kolb, C.E.; Worsnop, D.R.; Yourshaw, I.; Seinfeld, J.H.; Flagan, R.C.; Zhang, X.F.; Smith, K.A.; et al. Ambient aerosol sampling using the Aerodyne Aerosol Mass Spectrometer. *J. Geophys. Res. Atmos.* **2003**, *108*. [CrossRef]
17. Carlton, A.G.; Turpin, B.J. Particle partitioning potential of organic compounds is highest in the Eastern US and driven by anthropogenic water. *Atmos. Chem. Phys.* **2013**, *13*, 10203–10214. [CrossRef]
18. Faust, J.A.; Wong, J.P.S.; Lee, A.K.Y.; Abbatt, J.P.D. Role of Aerosol Liquid Water in Secondary Organic Aerosol Formation from Volatile Organic Compounds. *Environ. Sci. Technol.* **2017**, *51*, 1405–1413. [CrossRef]
19. Kampa, M.; Castanas, E. Human health effects of air pollution. *Environ. Pollut.* **2008**, *151*, 362–367. [CrossRef]
20. Nguyen, T.B.; Coggon, M.M.; Bates, K.H.; Zhang, X.; Schwantes, R.H.; Schilling, K.A.; Loza, C.L.; Flagan, R.C.; Wennberg, P.O.; Seinfeld, J.H. Organic aerosol formation from the reactive uptake of isoprene epoxydiols (IEPOX) onto non-acidified inorganic seeds. *Atmos. Chem. Phys.* **2014**, *14*, 3497–3510. [CrossRef]
21. Lin, Y.H.; Arashiro, M.; Clapp, P.W.; Cui, T.Q.; Sexton, K.G.; Vizuete, W.; Gold, A.; Jaspers, I.; Fry, R.C.; Surratt, J.D. Gene Expression Profiling in Human Lung Cells Exposed to Isoprene-Derived Secondary Organic Aerosol. *Environ. Sci. Technol.* **2017**, *51*, 8166–8175. [CrossRef] [PubMed]
22. Arashiro, M.; Lin, Y.H.; Zhang, Z.F.; Sexton, K.G.; Gold, A.; Jaspers, I.; Fry, R.C.; Surratt, J.D. Effect of secondary organic aerosol from isoprene-derived hydroxyhydroperoxides on the expression of oxidative stress response genes in human bronchial epithelial cells. *Environ. Sci. Process. Impacts* **2018**, *20*, 332–339. [CrossRef] [PubMed]
23. Verma, V.; Fang, T.; Xu, L.; Peltier, R.E.; Russell, A.G.; Ng, N.L.; Weber, R.J. Organic Aerosols Associated with the Generation of Reactive Oxygen Species (ROS) by Water-Soluble $PM_{2.5}$. *Environ. Sci. Technol.* **2015**, *49*, 4646–4656. [CrossRef] [PubMed]
24. Saffari, A.; Daher, N.; Shafer, M.M.; Schauer, J.J.; Sioutas, C. Global Perspective on the Oxidative Potential of Airborne Particulate Matter: A Synthesis of Research Findings. *Environ. Sci. Technol.* **2014**, *48*, 7576–7583. [CrossRef] [PubMed]
25. Fang, T.; Guo, H.; Verma, V.; Peltier, R.E.; Weber, R.J. $PM_{2.5}$ water-soluble elements in the southeastern United States: Automated analytical method development, spatiotemporal distributions, source apportionment, and implications for heath studies. *Atmos. Chem. Phys.* **2015**, *15*, 11667–11682. [CrossRef]
26. Sisler, J.; Malm, W. The relative importance of soluble aerosols to spacial and seasonal trends of impaired visibility in the United States. *Atmos. Environ.* **1994**, *28*, 851–862. [CrossRef]
27. Lippmann, M.; Ito, K.; Nadas, A.; Burnett, R.T. Association of particulate matter components with daily mortality and morbidity in urban populations. *Res. Rep. Health Eff. Inst.* **2000**, *95*, 5–72, discussion 73–82.
28. Tsai, F.C.; Apte, M.G.; Daisey, J.M. An exploratory analysis of the relationship between mortality and the chemical composition of airborne particulate matter. *Inhal. Toxicol.* **2000**, *12*, 121–135. [CrossRef]
29. Fairley, D. Daily mortality and air pollution in Santa Clara County, California: 1989–1996. *Environ. Health Perspect.* **1999**, *107*, 637–641. [CrossRef]
30. Mar, T.F.; Norris, G.A.; Koenig, J.Q.; Larson, T.V. Associations between air pollution and mortality in Phoenix, 1995–1997. *Environ. Health Perspect.* **2000**, *108*, 347–353. [CrossRef]

31. U.S. Environmental Protection Agency. *Quality Assurance Guidance Document 2.12*; Office of Air Quality Planning And Standards Air Quality Assessment Division: Research Triangle Park, NC, USA, 2016.
32. Eatough, D.J.; Wadsworth, A.; Eatough, D.A.; Crawford, J.W.; Hansen, L.D.; Lewis, E.A. A multiple-system, multichannel diffusion denuder sampler for the determination of fine particulate organic material in the atmosphere. *Atmos. Environ.* **1993**, *27*, 1213–1219. [CrossRef]
33. Turpin, B.J.; Huntzicker, J.J.; Hering, S.V. Investigation of organic aerosol sampling artifacts in the los angeles basin. *Atmos. Environ.* **1994**, *28*, 3061–3071.
34. El-Sayed, M.M.H.; Amenumey, D.; Hennigan, C.J. Drying-Induced Evaporation of Secondary Organic Aerosol during Summer. *Environ. Sci. Technol.* **2016**, *50*, 3626–3633. [CrossRef] [PubMed]
35. Zhang, X.; Liu, Z.; Hecobian, A.; Zheng, M.; Frank, N.H.; Edgerton, E.S.; Weber, R. Spatial and seasonal variations of fine particle water-soluble organic carbon (WSOC) over the southeastern United States: Implications for secondary organic aerosol formation. *J. Atmos. Chem. Phys.* **2012**, *12*, 6593–6607. [CrossRef]
36. Pöschl, U.; Shiraiwa, M. *Chemical Reviews*; American Chemical Society: Washington, DC, USA, 2015; pp. 4440–4475.
37. Christiansen, A.E.; Ghate, V.P.; Carlton, A.G. Aerosol Optical Thickness: Organic Composition, Associated Particle Water, and Aloft Extinction. *ACS Earth Space Chem.* **2019**, *3*, 403–412. [CrossRef]
38. Van Loy, M.; Bahadori, T.; Wyzga, R.; Edgerton, B.H.E. The aerosol research and inhalation epidemiology study (ARIES): $PM_{2.5}$ mass and aerosol component concentrations and sampler intercomparisons. *J. Air Waste Manag. Assoc.* **2000**, *50*, 1446–1458.
39. National Park Service. Improve—Interagency Monitoring of Protected Visual Environments. 2000. Available online: http://vista.cira.colostate.edu/improve/ (accessed on 13 July 2015).
40. Nguyen, T.K.V.; Capps, S.L.; Carlton, A.G. Decreasing Aerosol Water Is Consistent with OC Trends in the Southeast US. *Environ. Sci. Technol.* **2015**, *49*, 7843–7850. [CrossRef]
41. IMPROVE Public Archive. Available online: http://views.cira.colostate.edu/fed/DataWizard/Default.aspx (accessed on 13 July 2015).
42. Mesinger, F.; DiMego, G.; Kalnay, E.; Mitchell, K.; Shafran, P.C.; Ebisuzaki, W.; Jovic, D.; Woollen, J.; Rogers, E.; Berbery, E.H.; et al. North American regional reanalysis. *Bull. Am. Meteorol. Soc.* **2006**, *87*, 343. [CrossRef]
43. Fountoukis, C.; Nenes, A. ISORROPIA II: A computationally efficient thermodynamic equilibrium model for K^+-Ca^{2+}-Mg^{2+}-Nh_{4+}-Na^+-$SO_4{}^{2-}$-$NO_3{}^-$-Cl^--H_2O aerosols. *Atmos. Chem. Phys.* **2007**, *7*, 4639–4659. [CrossRef]
44. Moravek, A.; Murphy, J.G.; Hrdina, A.; Lin, J.C.; Pennell, C.; Franchin, A.; Middlebrook, A.M.; Fibiger, D.L.; Womack, C.C.; McDuffie, E.E.; et al. Wintertime spatial distribution of ammonia and its emission sources in the Great Salt Lake region. *Atmos. Chem. Phys.* **2019**, *19*, 15691–15709. [CrossRef]
45. Pye, H.O.T.; Nenes, A.; Alexander, B.; Ault, A.P.; Barth, M.C.; Clegg, S.L.; Collett, J.L.; Fahey, K.M.; Hennigan, C.J.; Herrmann, H.; et al. The Acidity of Atmospheric Particles and Clouds. *Atmos. Chem. Phys. Discuss.* **2019**. [CrossRef]
46. Petters, M.D.; Kreidenweis, S.M. A single parameter representation of hygroscopic growth and cloud condensation nucleus activity. *Atmos. Chem. Phys.* **2007**, *7*, 1961–1971.
47. Chan, M.N.; Choi, M.Y.; Ng, N.L.; Chan, C.K. Hygroscopicity of water-soluble organic compounds in atmospheric aerosols: Amino acids and biomass burning derived organic species. *Environ. Sci. Technol.* **2005**, *39*, 1555–1562. [CrossRef] [PubMed]
48. Burnett, R.T.; Stieb, D.; Brook, J.R.; Cakmak, S.; Dales, R.; Raizenne, M.; Vincent, R.; Dann, T. Associations between short-term changes in nitrogen dioxide and mortality in Canadian cities. *Arch. Environ. Health* **2004**, *59*, 228–236.
49. Nguyen, T.K.V.; Ghate, V.P.; Carlton, A.G. Reconciling satellite aerosol optical thickness and surface fine particle mass through aerosol liquid water. *Geophys. Res. Lett.* **2016**, *43*, 11903–11912. [CrossRef]
50. Farkas, C.M.; Moeller, M.D.; Felder, F.A.; Baker, K.R.; Rodgers, M.; Carlton, A.G. Temporalization of Peak Electric Generation Particulate Matter Emissions during High Energy Demand Days. *Environ. Sci. Technol.* **2015**, *49*, 4696–4704. [CrossRef]
51. Farkas, C.M.; Moeller, M.D.; Felder, F.A.; Henderson, B.H.; Carlton, A.G. High Electricity Demand in the Northeast US: PJM Reliability Network and Peaking Unit Impacts on Air Quality. *Environ. Sci. Technol.* **2016**, *50*, 8375–8384. [CrossRef]

52. El-Sayed, M.M.H.; Ortiz-Montalvo, D.L.; Hennigan, C.J. The effects of isoprene and NOx on secondary organic aerosols formed through reversible and irreversible uptake to aerosol water. *Atmos. Chem. Phys.* **2018**, *18*, 1171–1184. [CrossRef]
53. Youn, J.-S.; Wang, Z.; Wonaschütz, A.; Arellano, A.; Betterton, E.A.; Sorooshian, A. Evidence of aqueous secondary organic aerosol formation from biogenic emissions in the North American Sonoran Desert. *Geophys. Res. Lett.* **2013**, *40*, 3468–3472. [CrossRef]
54. Carlton, A.G.; Gouw, J.A.d.; Jimenez, J.L.; Ambrose, J.L.; Attwood, A.R.; Brown, S.; Baker, K.R.; Brock, C.; Cohen, R.C.; Edgerton, S.; et al. Synthesis of the Southeast Atmosphere Studies: Investigating fundamental atmospheric chemistry questions. *Bull. Am. Meteorol. Soc.* **2018**, *99*, 547–5567. [CrossRef]
55. Li, C.; Martin, R.V. Decadal Changes in Seasonal Variation of Atmospheric Haze over the Eastern United States: Connections with Anthropogenic Emissions and Implications for Aerosol Composition. *Environ. Sci. Technol. Lett.* **2018**, *5*, 413–418. [CrossRef]

atmosphere

MDPI

Article

Classification of Clouds Sampled at the Puy de Dôme Station (France) Based on Chemical Measurements and Air Mass History Matrices

Pascal Renard [1,*], Angelica Bianco [1,†], Jean-Luc Baray [1,2], Maxime Bridoux [3], Anne-Marie Delort [4] and Laurent Deguillaume [1,2,*]

1 Laboratoire de Météorologie Physique, UMR 6016, CNRS, Université Clermont Auvergne, 63178 Aubière, France; angelica.bianco@helsinki.fi (A.B.); j-luc.baray@uca.fr (J.-L.B.)
2 Observatoire de Physique du Globe de Clermont-Ferrand, UMS 833, CNRS, Université Clermont Auvergne, 63178 Aubière, France
3 Commissariat à L'énergie Atomique et aux Énergies Alternatives (CEA, DAM, DIF), 91297 Arpajon, France; maxime.bridoux@cea.fr
4 Université Clermont Auvergne, CNRS, SIGMA Clermont, Institut de Chimie de Clermont-Ferrand, F-63000 Clermont-Ferrand, France; a-marie.delort@uca.fr
* Correspondence: pascal.renard@uca.fr (P.R.); laurent.deguillaume@uca.fr (L.D.)
† Now at Institute for Atmospheric and Earth System Research/Physics, Faculty of Science, University of Helsinki, FI-00014 Helsinki, Finland.

Received: 12 June 2020; Accepted: 7 July 2020; Published: 10 July 2020

Abstract: A statistical analysis of 295 cloud samples collected at the Puy de Dôme station in France (PUY), covering the period 2001–2018, was conducted using principal component analysis (PCA), agglomerative hierarchical clustering (AHC), and partial least squares (PLS) regression. Our model classified the cloud water samples on the basis of their chemical concentrations and of the dynamical history of their air masses estimated with back-trajectory calculations. The statistical analysis split our dataset into two sets, i.e., the first set characterized by westerly air masses and marine characteristics, with high concentrations of sea salts and the second set having air masses originating from the northeastern sector and the "continental" zone, with high concentrations of potentially anthropogenic ions. It appears from our dataset that the influence of cloud microphysics remains minor at PUY as compared with the impact of the air mass history, i.e., physicochemical processes, such as multiphase reactivity.

Keywords: Puy de Dôme station (PUY); cloud chemistry and physics; air mass history; oceanic vs. continental influences; partial least squares (PLS) regression

1. Introduction

The chemical composition of cloud water has revealed the high complexity of this medium [1], resulting from the cloud scavenging of soluble gases, the dissolution of the soluble fraction of the aerosol acting as cloud condensation nuclei (CCN), and from aqueous phase reactions [2]. Additionally, recent studies have shown that microbial activity altered the cloud water chemical composition [3–5]. Therefore, cloud water is composed of a mixture of complex inorganic and organic compounds with strong oxidants which have been shown to drive aqueous phase transformations in the presence of solar radiation [6,7].

Studies devoted to the analysis of the chemical composition and to its variability have aimed to better understand several atmospheric physicochemical processes such as droplet activation and growth, production and consumption of chemical compounds, as well as transport and deposition by precipitation of pollutants. Many field studies have reported in cloud production of low-volatility products [8,9]. The production of strong acids such as sulfates has been highlighted in the cloud aqueous phase leading to acidification, a process that controls both the phase partitioning and chemical reaction rates [10] and is responsible for the widely known phenomenon of acid rain [11,12]. Oxidative cloud processing was suggested to form secondary organic aerosol (aqSOA) through functionalization of dissolved organic compounds [13]. These aqueous phase transformations alter particle properties in terms of mass, chemical composition, hygroscopicity, and oxidation state [14], also affecting their CCN ability. Variability of solute concentrations together with drop size distribution have been observed in many field studies, revealing variations of the CCN composition with particle size and the dependency of some processes (dissolution of soluble gases and condensational growth) on drop size [15–17]. In cloud scavenging efficiencies have also been investigated by simultaneously measuring cloud water concentrations and interstitial particulate or gaseous concentrations [18–20]. Deviations, up to a few orders of magnitude from Henry's law equilibrium, have been reported for carbonyl compounds with low effective water solubility [21,22] which suggested a more efficient scavenging of organic compounds by cloud water than expected. Deposition through precipitation of the scavenged chemical compounds removes large amounts of organic and inorganic pollutants from the atmosphere providing a significant contribution of nutrients (positive inputs) and pollutants (negative inputs) in various ecosystems [23,24].

Cloud water studies have been conducted over various continents [1] including Europe [25–28], Asia [7,29–31], North and South America [1,32–34], as well as in contrasted environments (polluted, marine, and remote). Significant developments have been implemented to enhance the collection efficiency of cloud collectors [35–37] and to better characterize the molecular composition by targeted or non-targeted methods, often using mass spectrometry [3,20,38]. Those studies mostly investigated the temporal variability of cloud chemical composition, as well as the transport of air masses and the physicochemical processes [30,33]. However, most of these field campaigns were performed over short-term periods due, in part, to the inherent difficulty of collecting clouds. A few sites have continuously collected cloud water over long-term periods such as the Puy de Dôme station in France (PUY) [39], Mt. Brocken in Germany [40], Whiteface Mountain in USA [32], Mt. Oyama in Japan [41] and Mt. Tai in China [31]. These mountain sites offer facilities to sample clouds under optimal conditions and to conserve quality homogeneity of chemical analysis on a long-term basis. These stations have a relatively high altitude, i.e., hundreds of meters above the surrounding plains, where cloud formation occurs, optimizing collection efficiency and sampling air masses from varied origins, throughout the year.

The present study aims at analyzing a long-term dataset of the chemical composition of cloud water samples at PUY. This remote site is influenced by long-range transport [42,43] and the proposed study has the objective to provide information on the physicochemical variability of air masses on a regional scale. For this, cloud water chemical composition is used to constrain a multivariate statistical analysis and propose a chemical classification of the sampled clouds. Then, this classification is combined to a numerical analysis using the CAT model (computing advection-interpolation of atmospheric parameters and trajectory tool) [39,44]. This model simulates the atmospheric transport of air masses and provides zone and sector matrices. Thereby, in regards to the basic back-trajectory methods [45], the CAT model brings additional information, and completes more robust statistical analyses, such as partial least squares (PLS) regressions between chemical and air mass history matrices. A specific point is addressed by the comparison of a previous cloud classification at PUY [46] and the identification of PUY's specificities, with similar studies [33,45,47–49]. Monitoring over 18 years also reveals some trends, both in terms of chemical concentrations and history of air masses that are discussed in this paper. Such a long-term monitoring avoids numerous statistical biases encountered in short-time field campaigns. The statistical information on the variability of chemical composition as a function of environmental factors (i.e., source regions) is also helpful for evaluating the extent of oceanic vs. continental influences and to define environmental contrasted scenarios for modeling purposes. Finally, the effect of cloud physics or air mass history on the chemical composition of clouds sampled at PUY is estimated.

2. Experiments

2.1. Cloud Sampling

Sampling was performed at PUY (45.7722° N, 2.9648° E), which belongs to several international networks as follows: EMEP (the European Monitoring and Evaluation Programme), GAW (Global Atmosphere Watch), and ACTRIS (Aerosols, Clouds, and Trace Gases Research Infrastructure). The observatory chalet is on top of a monogenetic volcano rising above the surrounding area with a height of 1465 m. PUY is part of the Chaîne des Puys, a north–south oriented chain of extinct volcanoes in the Massif Central (France), and to the west, an agricultural plain to the ocean (300 km apart). The urban area of Clermont-Ferrand and its surrounding suburbs (285,000 inhabitants) is situated 16 km east and 1000 m lower than the station. PUY is able to characterize air masses from various histories, coming from the boundary layer or in the free troposphere, varying as a function of the seasons and time of the day. The top of the mountain is frequently in cloudy conditions, on average 30% of the time per year, with higher occurrences during winter and autumn [50]. This makes PUY a reference site to study and sample cloud properties.

The cloud sampling dataset used in this study covered the period 2001–2018, with an average sampling time of 3 h and an average sampling volume of 75 mL. Non-precipitating cloud droplets were sampled using cloud collectors, as described previously for PUY cloud studies [46]. Cloud droplets, larger than 7 μm (cut-off diameter) [51], were collected by impaction onto a rectangular aluminum plate. Most of the time, droplets were collected directly as a liquid, and more rarely, they froze upon impaction (supercooled conditions). The water was transferred at room temperature, either directly or after a short melting period into glass vials. The aluminum collectors were cleaned and sterilized by autoclaving. Samples were collected in sterilized bottles and cloud water was filtered (0.20 μm nylon filter to eliminate microorganisms and particles). The majority of the sampled clouds resulted from frontal systems that mainly occurred during autumn, spring, and winter; these meteorological conditions enabled sampling clouds over long-time durations.

2.2. PuyCloud Database

The PuyCloud observation system deals with the monitoring of the biological, microphysical, and chemical properties of clouds. Biological and chemical analyses are performed in collaboration with the ICCF (Institute of Chemistry of Clermont-Ferrand). It is part of the French CO-PDD (Cézeaux-Aulnat-Opme-Puy De Dôme) multisite platform fully described by Baray et al. [39] in terms of instrumentations and data availability and widely employed [22,25,52–55].

The meteorological parameters that are monitored at PUY include the following: wind speed and direction, temperature, pressure, and relative humidity. Cloud microphysical properties, i.e., liquid water content (LWC) and effective droplets radius (r_e), are measured with a Gerber particle volume monitor, model 100 (PVM-100).

Physicochemical parameters are measured immediately after sampling (pH, conductivity, and redox potential). The concentrations of the major organic and inorganic ions (acetic, formic, succinic, malonic and oxalic acids, Ca^{2+}, K^+, Mg^{2+}, Na^+, NH_4^+, Cl^-, SO_4^{2-}, and NO_3^-) are measured by ion chromatography, using a DIONEX DX-320. The H_2O_2 and iron content, which are important parameters in the evaluation of the cloud oxidative capacity, are also determined. The spectrofluorimetric method based on the reactivity of p-hydroxyphenilacetic acid with horseradish peroxidase [53] was used to measure the concentration of hydrogen peroxide in cloud water. The Fe(II) concentration was measured by UV-visible spectroscopy at 562 nm, using the method developed by Stookey [56] based on the rapid complexation of iron with ferrozine. A detailed description of the physicochemical parameters and chemical analysis has been provided in Bianco et al. [6].

Between 2001 and 2018, 141 cloud events have been sampled at PUY, representing 295 individual samples (896 h of sampling), including 72 samples in spring, 21 samples in summer, 75 samples in fall, and 127 samples in winter. Table S1 indicates the physicochemical analysis performed for each cloud event. This cloud water chemical characterization was systematically performed for the last 20 years. Additional cloud water chemical and biological analysis have been developed during this last decade using targeted or global methods [3,22,25,52,54,57–61].

2.3. Dynamical Analysis

The trajectory approach is commonly used to identify source areas of air pollutants, based on conditional probability fields including back trajectory calculations, land cover, and meteorological data [62,63]. In the present work a dynamical analysis using the CAT model is performed to identify source areas of chemical compounds detected in cloud samples.

The CAT model is the recent evolution of the Lagrangian model LACYTRAJ [44]. CAT is a three-dimensional (3D) forward/backward kinematic trajectory code using initialization wind fields from the recent reanalysis ECMWF ERA-5 [64]. A cluster of starting back trajectory points is defined by the user and advected by the model using a bilinear interpolation for horizontal wind fields and time and a log-linear interpolation for vertical wind fields. The CAT model has already been used to determine the air masses arriving in PUY on the basis of calculations of two sets of 24 h back trajectories per day over a two-year period (2015–2016) [39].

In this study, sets of 45 back trajectories were calculated every hour during the cloud sampling in volume ± 0.1° in latitude and longitude. The vertical starting altitude of the back trajectories was deduced from the pressure measured at the Puy de Dôme summit considering the hydrostatic equilibrium. Trajectories were calculated between the summit and 50 m below (corresponding to 4 hPa) to take into account the ascent from the slopes of PUY of the air masses arriving below the observatory. The temporal resolution was 15 min and the total duration was 72 h.

The CAT model was initialized with ECMWF ERA-5 wind fields of any temporal and spatial resolution. For this work, wind fields were extracted every 3 h with a spatial resolution of 0.5° in latitude (55 km) and longitude (40 km), on 23 vertical pressure levels between 200 and 1000 hPa. CAT integrated a topography matrix at a resolution of around 10 km [65].

In addition to the wind parameters, the boundary layer height was also extracted from the ECMWF ERA-5 reanalysis in the same horizontal resolution, and spatially and temporally interpolated on all the trajectory points.

The trajectory calculation phase was followed by a dynamical characterization analysis phase. This phase consisted of flagging the cloud samples on the base of the results of the trajectories calculations.

The history of air masses was modeled by counting the number of trajectory points in each of the following nine geographic areas: north-northeast (NNE), east-northeast (ENE), east-southeast (ESE), south-southeast (SSE), south-southwest (SSW), west-southwest (WSW), west-northwest (WNW) and north-northwest (NNW), named "sector" hereafter, and one nearby area. The latter was defined because it was not possible to determine the origin for the closest points to PUY, in a radius of 0.5°, because of the spatial resolution of the wind fields. The percentage of points located over the sea and the continental surfaces was, then, determined using the topography file. If the altitude of the topography interpolated on each trajectory point is 0, this point is considered to be above the sea, and therefore corresponds to the "sea surface" zone. Otherwise it is the "continental surface" zone. Finally, we separated the continental and sea zones vertically, using the altitude of the atmospheric boundary layer height (ABLH) interpolated on the trajectory points (data summarized in Table S1, blue columns).

All of these characteristics were, then, compiled for each cloud sampling, providing a so-called "zone matrix" and a so-called "sector matrix". Thus, the matrices indicated, for each cloud sample, the distribution of the sectors or the zones crossed by their 72-hour backward trajectory. The relationship between the air mass history and the cloud composition was the subject of a statistical analysis, as described in Section 2.4.

2.4. Statistical Analysis

A principal component analysis (PCA) was performed using the concentrations of both organic and inorganic ions (Ca^{2+}, K^+, Mg^{2+}, Na^+, $NH_4{}^+$, Cl^-, $NO_3{}^-$, $SO_4{}^{2-}$, acetate, formate, malonate, oxalate, and succinate). The aim was to determine the most relevant variables to establish chemical categories (categories which would, then, be put in parallel with microphysical parameters or air mass history matrices). However, some chemical analyses were lacking and some samples presented missing values. Missing values were not replaced by the mean values to fully represent the variability of the dataset and to avoid overfitting [66]. Thus, samples, that were not fully characterized, were not considered in the statistical analysis.

Then, we performed numerous PCAs, and the maximum of information, gathered on the first two factors, was obtained with three ions with predominant marine sources (Cl^-, Mg^{2+}, and Na^+) and three ions with major sources from the continental surface (mostly anthropogenic ones) ($NH_4{}^+$, $SO_4{}^{2-}$, and $NO_3{}^-$). Hence, by keeping six inorganic ions, as similar studies [33,46,47], we fulfilled the best balance between samples and variables (i.e., increasing the number of variables means decreasing the number of samples).

The PCA type that was used during the computations was the Spearman's correlations (more appropriate when running the PCA on variables with different distributions).

Then, we performed agglomerative hierarchical clustering (AHC), an iterative classification method, the aim of which was to make up homogeneous groups of objects (categories) on the basis of their description by a set of variables (chemical variables, herein) describing the dissimilarity between the objects (cloud events, herein). The AHC produced a dendrogram which showed the progressive grouping of the data. To calculate the dissimilarity between samples, we applied the common Ward's agglomeration method (which minimized the within-group inertia) using Euclidean distance. The data were centered-reduced, to avoid variables with strong variance which unduly weighed on the results. The truncation level was automatically defined on the base of the entropy, and therefore the number of categories to retain.

Then, a PLS regression was performed to establish the correlations between the chemical parameters and the air masses history parameters. The Mann–Whitney and Kruskal–Wallis nonparametric tests were carried out to validate significant differences between two and among several data groups, respectively. Two air mass categories were declared to be different when the probability for the groups to have identical data distribution was lower than 5% (p-value < 0.05). These tests were chosen because the population from which the sample was extracted did not follow a normal distribution, according to the Shapiro–Wilk normality test.

This statistical analysis was performed using Excel XLSTAT software [67].

3. Results and Discussion

The multivariate statistical analysis was performed on 295 cloud samples collected at PUY, starting with PCA and AHC, in order to classify them according to their chemical composition. Then, these results were compared to the previous PUY study [46]. Then, we investigated, by using PLS regression, the relationships among these chemical data and the matrices provided by the CAT model, both on the zones ("sea and continental surfaces") and on the cardinal sectors crossed by the air masses. Finally, we compared the respective influences of the air mass history and microphysics on the chemical composition of clouds.

3.1. Clusterization of Cloud Waters at PUY

Data relative to Cl^-, Mg^{2+}, Na^+, NH_4^+, NO_3^-, and SO_4^{2-}, presented in Table S1, were analyzed by AHC and PCA to obtain categories based on ion concentration dissimilarities.

3.1.1. Chemical Categories

AHC was used to categorize cloud samples based on the long-term monitoring of their chemical composition. The AHC algorithm successfully grouped all the observations with a satisfactory cophenetic correlation (correlation coefficient between the dissimilarity and the Euclidean distance matrices) of 0.619 (Figure 1). Indeed, the closer the correlation to 1, the better the quality of the clustering. The dotted line in Figure 1 represents the degree of truncation (dissimilarity = 91.16) of the dendrogram used for creating categories and was automatically chosen based on the entropy level. Given the small difference in dissimilarity (Figure 1) between the light blue category (dissimilarity = 88.91) and the aggregate of the yellow and red categories (dissimilarity = 93.41), the AHC could be almost as robust with three or five categories. Nevertheless, these four AHC categories are consistent with our previous study [46]. Categories 1, 2, 3, and 4 consist of 113, 31, 55 and 9 clouds, respectively.

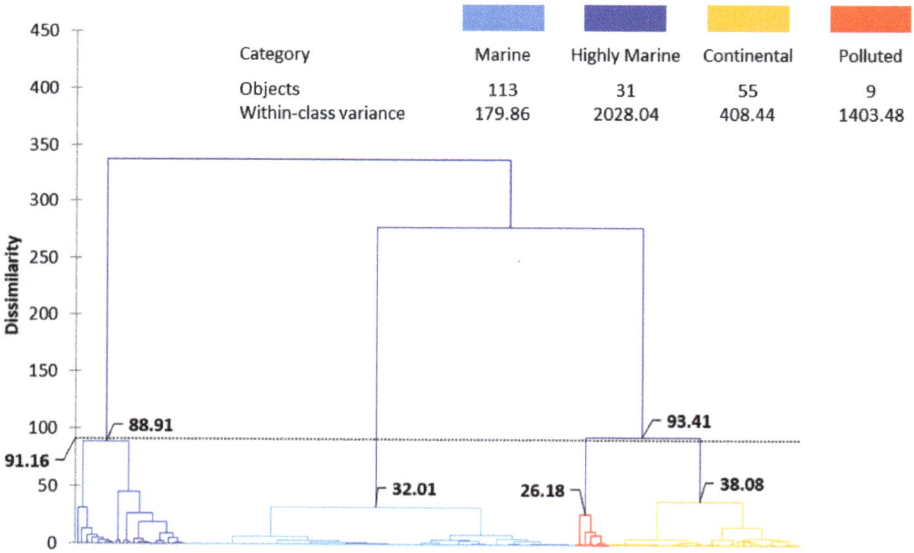

Figure 1. Dendrogram representing the agglomerative hierarchical clustering (AHC) based on dissimilarities using the Ward's method on 6 inorganic ion concentrations. The 208 cloud samples (without chemical missing values) were assigned to one of four automatically established categories (dissimilarities values displayed in bold). The six ions are the same as those used for principal component analysis (PCA).

The ACH profile plot (Figure 2), represents the four categories determined from the six main inorganic ions (Cl^-, Mg^{2+}, Na^+, NH_4^+, NO_3^-, and SO_4^{2-}). The light blue category with low ion concentrations is named "Marine" according to its air mass history (i.e., the time spent by this air mass above the "sea surface", detailed in Section 3.2). This category is the most homogeneous. This is confirmed by its significantly lower within-class variance (179.86), as shown in Figure 1. The "marine" category is also the main category (113 objects), which is consistent with the remoteness of the PUY. The dark blue category is characterized by high concentrations of Na^+, Cl^-, and Mg^{2+} and its air mass history, and thus called "highly marine". PUY is located more than 300 km from the Atlantic shore.

Nevertheless, at a synoptic scale, the air masses are mainly transported from the Ocean to PUY with no relief between (as confirmed hereafter by the CAT model). Hence, this category, with 31 objects, would appear to be counterintuitively modest. This suggests that some western clouds (which could have been classified as "highly marine") have either precipitated or become diluted (increase in liquid water content), thereby decreasing concentration. Then, these western clouds are classified "marine", hence, the importance of this category (i.e., a category with a marine history, but without salt).

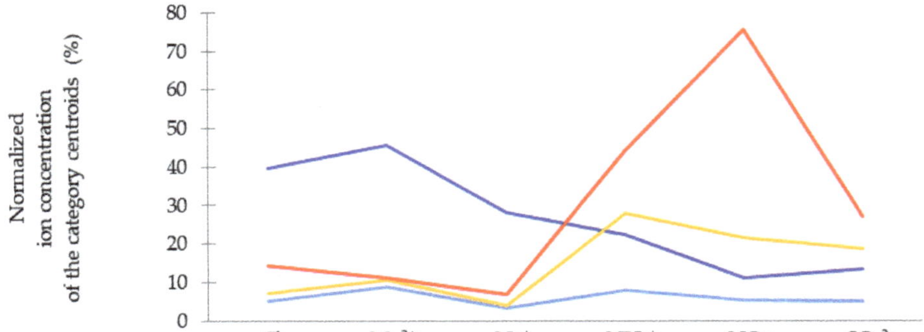

	Cl^-	Mg^{2+}	Na^+	NH_4^+	NO_3^-	SO_4^{2-}
Marine	4.96	8.90	3.38	7.95	5.24	4.99
Highly marine	39.72	45.67	28.22	22.34	11.23	13.46
Continental	7.12	10.42	3.99	27.69	21.37	18.68
Polluted	14.17	11.09	6.85	44.31	75.53	27.08

Figure 2. Profile plot established by the AHC from the six main inorganic ions (Na^+, Cl^-, Mg^{2+}, SO_4^{2-}, NH_4^+, and NO_3^-). The Y axis displays the normalized $(([Cl^-] - [Cl^-]_{min})/([Cl^-]_{max} - [Cl^-]_{min}))$ ion concentrations of the category centroids.

In red, the smallest category (nine objects), referred to as "polluted" in Figure 1 displays peak concentrations for SO_4^{2-}, NH_4^+, and NO_3^-, suggesting the air mass passed over an urbanized area. Below these maxima, in yellow, the "continental" category with 55 objects stands out. It should be noted, with only nine objects, the polluted category is statistically less robust, and could have been merged with the "continental" category (see dissimilarities in Figure 1), and regarded as the extreme SO_4^{2-}, NH_4^+, and NO_3^- values of the category. Conversely, the "highly marine" category could have been split (see dissimilarities in Figure 1), according to their SO_4^{2-} concentration (not shown).

Because the computed *p*-value in the Kruskal–Wallis test (Figure 3) is lower than the significance level alpha = 0.05, the distribution of ions (Cl^-, Mg^{2+}, Na^+, NH_4^+, NO_3^-, and SO_4^{2-}) concentration can be accepted as significantly different between each category. The samples do not come from the same population. We observe, in particular, high sea salts concentrations (Cl^-, Mg^{2+}, and Na^+) for both "marine" and "highly marine" categories, and high concentrations of potentially anthropogenic ions (NH_4^+, NO_3^-, and SO_4^{2-}) [68–72] for both "polluted" and "continental" categories (Table S2).

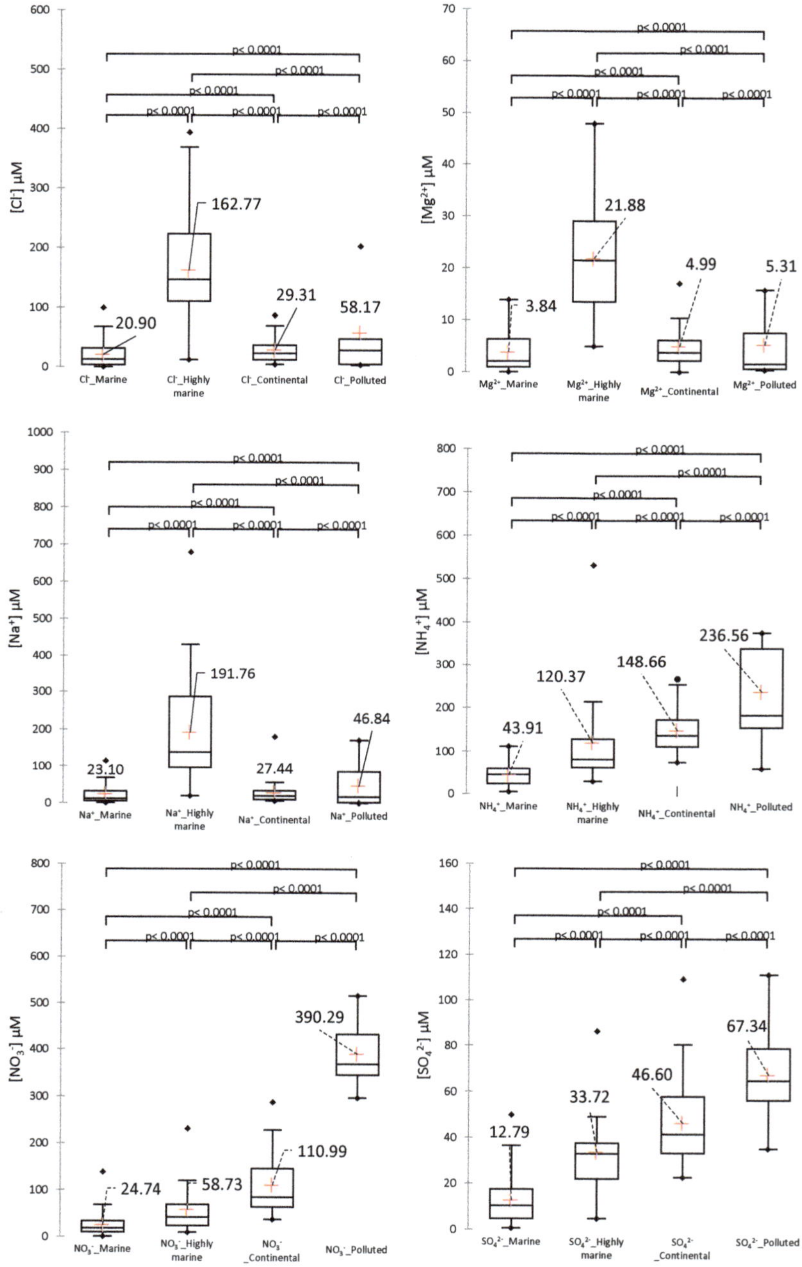

Figure 3. Distribution of inorganic ions (Cl^-, Mg^{2+}, Na^+, $NH_4{}^+$, NO_3, and $SO_4{}^{2-}$) of the cloud waters sampled at PUY for each air mass category (marine, highly marine, continental, and polluted). The number of analyzed samples is 208 (samples with missing data were removed). One box plot per category is displayed for each ion. The mean values are displayed as red crosses. The central horizontal bars are the medians. The lower and upper limits of the box are the first and third quartiles, respectively. The ends of whiskers are 10th and 90th percentiles. Black diamonds are minimum and maximum for each species. The box plot's horizontal width has no statistical meaning. Statistical differences (Kruskal–Wallis test; p value < 0.05) between groups are indicated above box plots.

3.1.2. Variable Validation

A PCA was computed on a Spearman correlation matrix using the concentrations of the ions (Cl^-, Mg^{2+}, Na^+, NH_4^+, NO_3^-, and SO_4^{2-}). The PCA correlation circle (Figure 4a) provides evidence that Cl^-, Mg^{2+}, and Na^+ are strongly correlated (see correlation matrix in Table S3, r(Na^+, Cl^-) = 0.82, r(Mg^{2+}, Cl^-) = 0.77, and r(Na^+, Mg^{2+}) = 0.77); as well as NH_4^+, NO_3^-, and SO_4^{2-} (r(NO_3^-, NH_4^+) = 0.77; r(NO_3^-, SO_4^{2-}) = 0.75; and r(NH_4^-, SO_4^{2-} = 0.78); while these two sets are practically uncorrelated, except Cl^- and SO_4^{2-} (r(Cl^-, SO_4^{2-}) = 0.49), suggesting the presence of anthropogenic chlorine and sea salt sulphate.

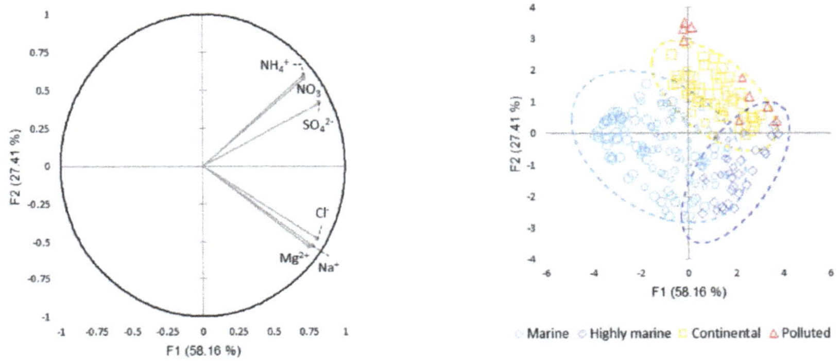

(a) (b)

Figure 4. Principal component analysis (PCA) on a Spearman correlation chemical matrix. (**a**) Correlation circle and projection of the 6 ion concentrations; (**b**) Two-dimensional map of the colored observations according to the AHC category. The XLstat software automatically displayed confidence ellipses (interval 95%) around AHC categories, and resized points with squared cosines of the observations (i.e., the larger the point, the more it is related to a factor, F1 or F2).

In this PCA (Figure 4), the first two factors represent 85.57% of the initial variability of the data; the PCA is robust, with no information hidden in the next four factors (see squared cosines of the variables in Table S4). The horizontal axis (F1) is linked to the total ion concentration and represents 58.16% of the information, while the vertical axis (F2: 27.4%) is linked to the concentrations of NH_4^+, NO_3^-, and SO_4^{2-} in positive, and Cl^-, Mg^{2+}, and Na^+ in negative. The PCA is consistent with the AHC. Coherently, in Figure 4b, the AHC "marine" category stands out on the left (F1 < 0) of the chart, the "highly marine" category at the bottom right (F1 > 0 and F2 < 0), the "continental" and the "polluted" categories at the top right (F1 > 0 and F2 > 0).

3.1.3. Evolution Since the 2001–2011 Study

In this study, the statistical analysis evolves as compared to our previous work. First, the AHC is performed with a larger number of samples (208 versus 134) and variables considered for the statistical analysis are different, i.e., pH is not taken into account and Mg^{2+} is added to the variables, as explained above. We removed cloud events with missing values. The ACP Spearman's correlations replaced Pearson's. However, the distribution of categories is fairly unchanged; among the 208 cloud events used in the AHC, 164 (78.8%) were clustered in a category similarly named in the 2001–2011 study [46] (see Table S1).

The samples with high Cl^-, Mg^{2+}, and Na^+ concentrations are still gathered in the so-called "highly marine" (HM) category. However, the present HM category is an expanded version of the former "highly marine" ($HM_{\_01\text{–}11}$) category, with lower mean concentrations of Cl^- and Na^+ ($[Na^+]_{HM}$ = 192 μM vs. $[Na^+]_{HM\_01\text{–}11}$ = 311 μM and $[Cl^-]_{HM}$ = 163 μM vs. $[Cl^-]_{HM\_01\text{–}11}$ = 232 μM). The average ratio Cl^-/Na^+ of this updated "HM" category is higher 1.22 vs. 1.06. Among the 31 cloud events in HM, only 10 were clustered in "$HM_{\_01-11}$".

The "marine" category is barely larger in percentage than the former one ($marine_{\_01-11}$), with lower Na^+, Cl^-, and SO_4^{2-} concentrations ($[Na^+]_{Marine}$ = 23.3 μM vs. $[Na^+]_{Marine\_01-11}$ = 32 μM, $[Cl^-]_{Marine}$ = 20.5 μM vs. $[Cl^-]_{Marine\_01-11}$ =30 μM, and $[SO_4^{2-}]_{Marine}$ = 12.8 μM vs. $[SO_4^{2-}]_{Marine\_01-11}$ = 28 μM). NH_4^+ and NO_3^- are equivalent. Among the 113 cloud events in "marine", 111 are clustered in "$marine_{\_01-11}$".

Conversely, the "continental" category, is slightly smaller in percentage than the former one ($continental_{\_01-11}$), 26% of the samples vs. 34%, although the mean SO_4^{2-} concentration increases ($[SO_4^{2-}]_{Continental}$ = 46.6 μM vs. $[SO_4^{2-}]_{Continental\_01-11}$ = 94 μM). The other concentrations remain almost unchanged. Among the 35 cloud events in "continental", 35 were clustered in "$continental_{\_01-11}$".

In both studies, 4% of cloud samples are in the "polluted" category, while the mean ion concentrations are markedly lower. Among the nine cloud events in "polluted", eight are clustered in "$polluted_{\_01-11}$".

In summary, the "marine" category slightly increases in percentage, as the mean ion concentration of the "continental" and "polluted" categories dwindle, in particular for the anthropogenic ions. "Highly marine" is the category that has expanded the most. This trend is not fully explained by the minor statistical processing adjustments (see Section 2.4). We compared (not shown) the two methods on the first 2001–2011 dataset, without observing any significant difference.

We performed a Mann–Whitney test (Figure 5) on the clouds sampled from 2001 to 2011 (period covered by our previous work [46]) and since then. It appears that NH_4^+, NO_3^-, and SO_4^{2-} concentrations are significantly lower on this second period (2012–2018), i.e., ($[NH_4^+]_{01-11}$ = 96 μM vs. $[NH_4^+]_{12-18}$ = 78 μM, $[NO_3^-]_{01-11}$ = 76 μM vs. $[NO_3^-]_{12-18}$ = 44 μM, and $[SO_4^{2-}]_{01-11}$ = 31 μM vs. $[SO_4^{2-}]_{12-18}$ = 27 μM (p-values are 0.027, 0.0009, and 0.009, respectively). The concentration of sea salts does not evolve significantly, but the changes on anthropogenic classes (mentioned above) drive the changes observed in the "marine" and "highly marine" categories. Category terminology will receive additional justifications in Section 3.2.

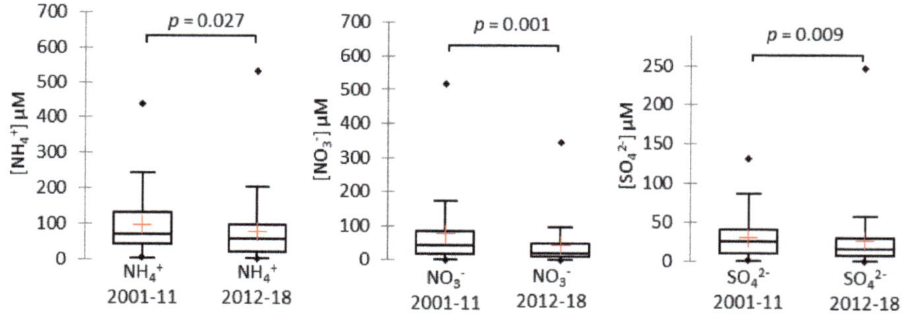

Figure 5. Mann–Whitney nonparametric tests on 154 clouds sampled from 2001 to 2011, the period covered by the previous study [46], and 88 clouds sampled from 2012 to 2018. We compare, for both periods, NH_4^+, NO_3^-, and SO_4^{2-}concentrations. The p-values of all pairwise comparisons are significant at level alpha = 0.05.

3.2. Influence of Air Mass History at PUY

This section is devoted to the correlation between the concentration of the inorganic ions and the air mass history. During their atmospheric transports, the air masses received chemical species under various forms (gases and particles) from various sources. This strongly depended on the altitude of the air masses. During the transport, chemicals could also undergo multiphasic chemical transformations, as well as dry or wet deposition. The objective, here, is to evaluate the effect of the history of air masses on the chemical composition of clouds. To this end, PLS regressions are performed and the results are validated with nonparametric tests (Kruskal–Wallis and Mann–Whitney tests).

As described in Section 2.4, the CAT model provides two matrices. The "zone matrix" contains information about the time spent by air masses over "continental surface" or "sea surface", in the atmospheric boundary layer (<ABLH) or in the free troposphere (>ABLH). The "sector matrix" contains information about the time spent by air masses in the eight forty-five degrees sectors (NNE, ENE, ESE, SSE, SSW, WSW, WNW and NNW; see Figure S1c). Figure 6a represents the distribution of these parameters for all the cloud events. Despite the distance from the coast (300 Km), the strong maritime influence at PUY is obvious (Figure 6a). Over a 72-hour backward trajectory, on average, an air mass spends almost two days over the "sea surface". Coherently, PUY is characterized by prevailing strong west and north winds (WSW, WNW, NNW, and NNE), with average percentages of time spent over these four main sectors of 23, 44, 14, and 12%, respectively (Figure 6b).

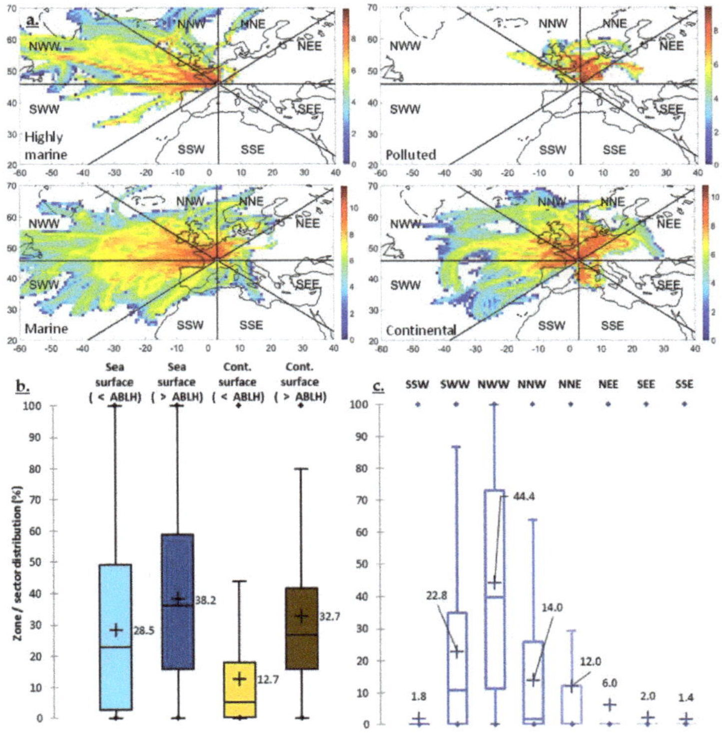

Figure 6. (**a**) Natural logarithm of the number of CAT (computing advection-interpolation of atmospheric parameters and trajectory tool) back trajectories points arriving at the summit of the PUY station per square of 0.8° size for "highly marine" (top left), "marine" (bottom left), "polluted" (top right), and "continental" (bottom right) categories. The black lines separate the different sectors. Distributions of the parameters evaluated by the CAT model for all the 295 cloud events; (**b**) Percentage of the time spent over the "sea surface" (blue) and "continental surface" (brown), in pale blue and pale brown below the atmospheric boundary layer height (<ABLH) and dark above in the free troposphere (>ABLH); (**c**) Percentage of the time spent in the 8 forty-five degrees sectors (NNE, ENE, ESE, SSE, SSW, WSW, WNW and NNW). One box plot per zone/sector is displayed. The black crosses correspond to the means. The central horizontal bars are the medians. The lower and upper limits of the box are the first and third quartiles, respectively. The ends of whiskers are 10th and 90th percentiles. Black diamonds are minimum and maximum for each species.

To perform the PLS analysis (Figure 7), the matrix of the explanatory variables (the "Xs") is composed of the "sector matrix" and the "zone matrix". The matrix of the dependent variables (the "Ys") is the chemical matrix. As explained in Section 2.3, we restricted our statistical analysis to the concentration of six chemical compounds to avoid excessive loss of information and overfitting in the statistical analyses.

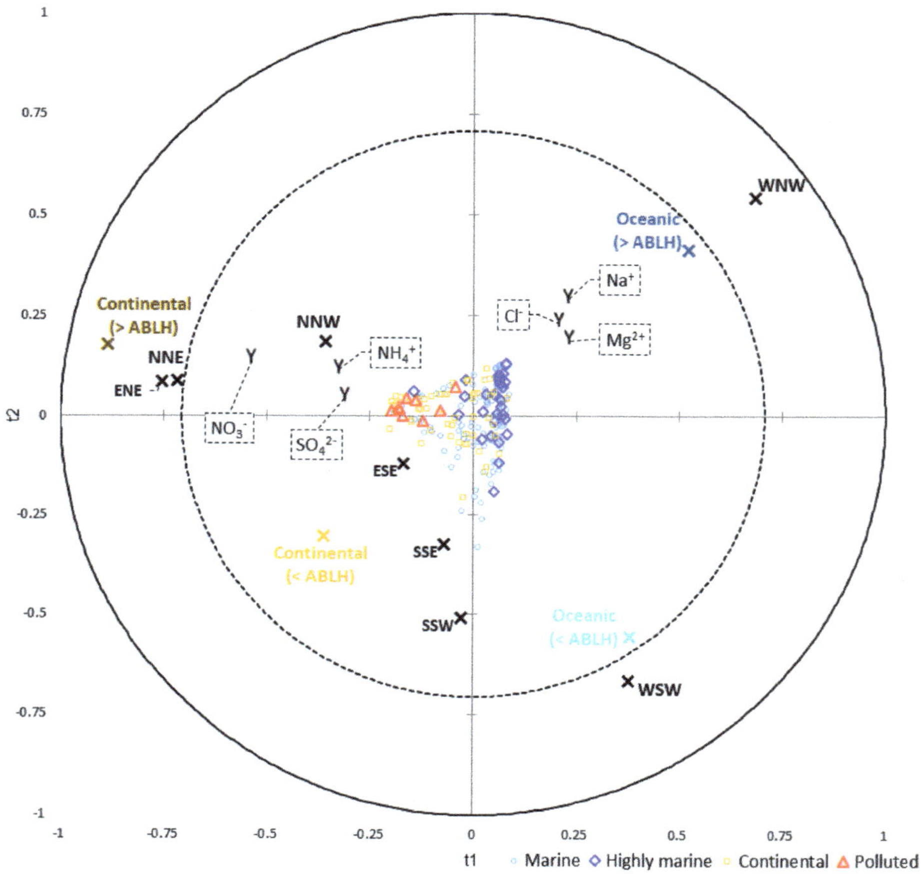

Figure 7. Partial least squares (PLS) chart with t component on axes t1 and t2. The correlations map superimposes the "Xs", the "Ys" and the cloud events. The dependent variables from the chemical matrix are symbolized by a black "Y"; the explanatory variables from the "sector matrix" by a black "X"; and from the "zone matrix" by a blue, brown, light or dark "X". The 208 cloud events are gathered by AHC category (red circle, "marine"; dark blue diamond, "highly marine"; yellow square, "continental"; and red triangle, "polluted").

The index of the predictive quality of the models is quite low ($Q^2 = 0.1$, ideally it should be close to 1) suggesting weak correlations. It is well known that cloud composition depends on many other parameters than the chosen explanatory variables, related to the air mass history calculated by the model. Indeed, cloud chemical composition depends foremost on local microphysics [17,37,73], proximity to sources [33,48,74,75], biological activity [4,5,61], seasonal cycles [30,76–78], and diurnal cycles [79].

Figure 7 displays numerous intricacies between chemical parameters and the air mass history. First, some zone variables are weakly correlated to some sector variables (cf. PLS correlation matrix in Table S5), "sea surface" (>ABLH) with WNW), "sea surface" (<ABLH) with WSW), "continental surface" (>ABLH) with ENE (too few observations to be interpretable on the graph) and more robustly, "continental surface" (>ABLH) with ENE (R = 0.7).

The "polluted" category in red (Figure 7) and, to a lesser extent, the "continental" category in yellow are on the left of the display, toward the NNE sector and the "continental surface" (>ABLH) zone. The "highly marine" category in dark blue and, to a lesser extent, the "marine" category in light blue are drawn toward the WNW/WSW sectors and the "sea surface" (>ABLH) zone.

We performed an AHC on the "sector matrix" and obtained three clusters. Then, we reran the previous PLS. The simplified correlation matrix (Table 1) highlights the link between "sea surface" zones west sector and "continental surface" zones and northeast sector. We do not keep this clusterization in the main PLS to avoid a loss of information.

Table 1. PLS correlation matrix between clustered sector variables and zone variables. Highest correlation displayed in red and highest anti-correlation in blue.

Variables	"Sea Surface" (<ABLH)	"Sea Surface" (>ABLH)	"Continental Surface" (<ABLH)	"Continental Surface" (>ABLH)
WSW_WNW	0.35	0.38	−0.27	−0.72
NNW_NNE_ENE	−0.33	−0.44	0.31	0.76
ESE_SSE_SSW	−0.03	−0.02	−0.04	0.07

The simplified correlation matrix (Table 2) displays weak correlations. However, the link between sea salts (Cl^-, Mg^{2+}, and Na^+) and both the "sea surface" (>ABLH) zone and the WSW/WNW clustered sectors is noticeable. The same applies to ions of potentially anthropogenic origin (NH_4^+, NO_3^-, and SO_4^{2-}) and both the "continental surface" (>ABLH) zone and the NNW/NNE/ENE sector. For both marine and continental ions, the correlations are higher above the atmospheric boundary layer height (>ABLH), confirming PUY is surely influenced by long-range transport [42,43].

Table 2. PLS correlation matrix between chemical variables and both zone and clustered sector variables. Highest correlation displayed in dark red and highest anti-correlation in dark blue.

Variables	"Sea Surface" (<ABLH)	"Sea Surface" (>ABLH)	"Continental Surface" (<ABLH)	"Continental Surface" (>ABLH)	WSW WNW	NNW NNE ENE	ESE SSE SSW
Cl^-	0.03	0.14	−0.12	−0.16	0.16	−0.13	−0.10
Mg^{2+}	0.06	0.14	−0.13	−0.19	0.18	−0.15	−0.10
Na^+	−0.09	0.25	−0.16	−0.16	0.20	−0.18	−0.08
NH_4^+	−0.28	−0.03	0.03	0.31	−0.30	0.30	0.01
NO_3^-	−0.25	−0.26	0.15	0.51	−0.46	0.51	−0.04
SO_4^{2-}	−0.18	−0.06	0.12	0.21	−0.35	0.30	0.09

In order to statistically validate these observations, we performed the Kruskal–Wallis test and compared the category distribution within each zone (Figure S1a) and main sectors (Figure S1b). As the computed p-values are lower than the significance level alpha = 0.05, we accept that the main sectors (WSW, WNW, NNW, NNE, and ENE) and the zones ("sea surface" (>ABLH), "sea surface" (<ABLH), and "continental surface" (>ABLH)) are significantly different for each category. The samples do not come from the same population. Only the p-value of "continental surface" (>ABLH) is greater than the significance level alpha = 0.05 (p-value = 0.062). The difference between the categories according to the sector distribution can also be observed on the map (Figure S1c) provided by the CAT model. The history of air masses significantly influences the chemical composition of clouds.

3.3. Influence of Cloud Microphysics at PUY

The air mass history can influence solute concentration by scavenging aerosol particles and gaseous species (as discussed in Section 3.2). This strongly depends on the CCN concentration related to the physicochemical properties of aerosol particles (size distribution and chemical composition) and on the gas phase chemical composition and corresponding phase equilibria. Microphysical cloud conditions such as liquid water content (LWC) and effective droplets radius (r_e) can also perturb solution concentration variability, as well as chemical reactions occurring within cloud waters. This section is devoted to the possible relationships between LWC and chemical variables.

For this, a PLS analysis was performed with the LWC and the r_e, as the matrix of the explanatory variables (the "Xs"), and the chemical matrix (the "Ys"). The PLS chart is presented in Figure 8, and Table S6 reports the correlation matrix between these variables. There are weak correlations between LWC and ion concentrations. The strongest anti-correlation is between NH_4^+ and r_e, i.e., $r(NH_4^+, r_e) = -0.37$. This analysis clearly demonstrates that at PUY, microphysical properties of the sampled clouds are almost not correlated with their chemical composition. This could be explained by the type of clouds that are collected; the majority are frontal clouds which were formed well before their arrival at the top of the mountain and which present a low variability in their microphysical properties.

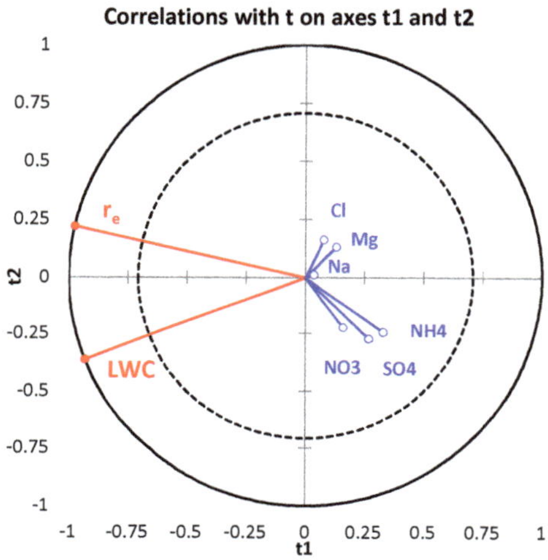

Figure 8. Chemistry/Microphysics PLS chart of cloud samples with t component on axes t1 and t2. Correlations map allows superimposing the "Xs" and the "Ys" (observations are removed for clarity). PLS performed on 73 cloud events. The dependent variables from the chemical matrix are displayed in blue, the explanatory variables from the microphysics matrix in red.

A supplementary analysis (PCA) has also been performed (Table S7), demonstrating that correlations between microphysical variables and air mass history parameters (Section 3.2) are negligible. Thus, the influence of air mass history on the chemical composition of clouds cannot be attributed to the variability of microphysical parameters. In other words, there is an influence of microphysics, but it is statistically identical, whatever the zones or the sectors crossed by the air mass.

To remove any influence of LWC variation, cloud water loadings (CWLs) are commonly calculated to evaluate the solute content per volume of air. The statistical analysis, in this study, were conducted on solute concentrations in cloud water to get more robust results, because microphysical parameters are not always available, especially under winter conditions. However, LWC and r_e variations are not highly variable (Figure S2) at PUY, and CWL patterns resemble those of solute concentrations in cloud waters (Figure S3). This suggests that for clouds sampled at PUY, the air mass history can better explain the variability of cloud water solute concentrations than LWC variations.

A decrease of the solute concentrations from continental origin (NO_3^-, SO_4^{2-}, and NH_4^+) was observed between the periods 2001–2011 and 2012–2018, as mentioned in Section 3.1.3. In addition, a low decrease of its mean value (from 0.31 to 0.27 g·m^{-3}) suggests that the CWLs for these species also significantly decreased. This trend could be explained by the aerological evolution highlighted by the CAT model and requires further investigation.

Previous field studies have investigated the dependency of cloud chemical composition with microphysical parameters [16,80–82]. It has been shown, for sites more exposed to anthropogenic emissions, that LWC could modulate solute concentrations. For example, clouds freshly formed by orography can have their chemical composition modulated by cloud microphysics [7,19,73]. Anthropic mean cloud water concentrations at PUY are notably low ($[NH_4^+]$ = 91 µM, $[NO_3^-]$ = 70 µM, and $[SO_4^{2-}]$ = 27 µM, see Table S2) as compared with the literature data [19]. Polluted events are exceptionally observed at PUY and most likely originate from afar (northeast France, several hundred kilometers away). If we consider the synoptic scale, we should see higher concentrations of NH_4^+, NO_3^-, and SO_4^{2-}; however, these ions are involved in chemical and photochemical reactions [83–86]. Hence, more information on the chemical aging of the air masses is needed (chemical characterization in progress). Moreover, such a long-term monitoring, with varied air masses, smooths the microphysics (LWC and r_e) influence (Figure S2). Cloud water is a complex matrix resulting from the interaction of many factors. Nevertheless, it appears that the air mass history, despite reduced correlations, remains the prevailing parameter, with either western and oceanic clouds or northeastern anthropogenic clouds.

4. Conclusions

In this study, statistical analyses (AHC and PCA) were carried out on 208 cloud samples collected at the Puy de Dôme station (France) between 2001 and 2018, which resulted in clustering the cloud samples according to their chemical properties (concentrations of inorganic ions from marine and continental origins) into four categories as follows: "highly marine", "marine", "continental", and "polluted". Despite an evolution of the statistical treatment to classify the clouds samples, this work confirms those established in a previous study by Deguillaume et al. [46] for clouds sampled between 2001 and 2011. A change between the relative proportions of categories is however noticed and attributed to a significant decrease in the NH_4^+, NO_3^-, and SO_4^{2-} concentrations during the second period (2012–2018) of cloud sampling.

CAT models the history of the air masses arriving at PUY, providing for each air mass the time spent above the eight cardinal sectors and above continental or sea surfaces. The CAT model specifies whether the air mass is in a free troposphere or in an atmospheric boundary layer. From these in silico zone and sector matrices and in situ chemical characteristics, PLS analysis highlights two main relationships between air mass origins and ion concentrations. A type of air mass comes predominantly from western sectors and from the "sea surface" (> ABLH) zone, with the highest concentrations of sea salts (Cl^-, Mg^{2+}, and Na^+). A total of 31 cloud samples are gathered in the "highly marine" AHC category, which are characteristic of this air mass. Slightly linked to the latter, the "marine" AHC category, which is named for its air mass history and its low ion concentrations, is the most important (113 cloud samples) and the most "homogeneous". The second main air mass type arrives from the northeast sector and from the "continental surface" (> ABLH) zone, with the highest concentrations of potentially anthropogenic ions (NH_4^+, NO_3^-, and SO_4^{2-}). Only nine cloud samples are grouped in the "polluted" AHC category, characteristic of this air mass. With less extreme values and 55 cloud samples, the "continental" category represents the body of this set.

Finally, the influence of cloud microphysical properties (LWC and r_e) on the cloud water composition is investigated using PLS analysis in a similar way. This indicates no robust statistical correlations between cloud microphysics and cloud water chemical composition. This suggests that cloud chemical composition at PUY is influenced by air mass history which includes several physicochemical processes (CCN physical and chemical processes, mass transfer of soluble species, multiphase reactivity, etc.).

Clearly, this study highlights parameters that could drive the chemical composition of clouds at PUY; this statement cannot be generalized to other observation sites presenting different environmental scenarios. However, in a remote site, it appears that without major and immediate urban or marine influence, an air mass coming from the ocean or from a polluted area would be observed more or less loaded, according to complex biophysicochemical processes. In addition, much of the oceanic influence (i.e., Cl^-, Mg^{2+}, and Na^+ concentrations) seems to decrease quickly (78% of the clouds coming from the ocean appear "cleaned"), and much of the anthropic influence seems more persistent (NH_4^+, NO_3, and SO_4^{2-} concentrations) which remain significant.

The PUY site is a reference European station for the study of gases, aerosols, and clouds. International field campaigns have been conducted there in the past and future campaigns would especially target cloud biophysicochemical processes. Cloud waters collected at PUY for various air mass histories also serve for laboratory investigations that consider the following: (1) characterizing the complex chemical composition and its environmental variability by innovative analytical methods, and (2) quantifying photochemical and biological transformations occurring in this complex liquid medium. Cloud field investigations performed at PUY also help to build relevant chemical scenarios that help to better constrain cloud chemistry models [87]. For the dynamical frame, the CAT model makes it possible to give an overview of the air mass history; this helps to constrain cloud chemistry models but also makes it possible to compare the PUY station to other observatories where cloud studies are conducted.

Supplementary Materials: The following are available online at http://www.mdpi.com/2073-4433/11/7/732/s1, Figure S1: Category distribution within each zone (a) and sector (b), Figure S2: LWC distribution at PUY, Figure S3: Comparison of normalized concentrations and normalized CWLs, Table S1: PuyCloud data ("TableS1.xlsm"), Table S2: Ion concentrations of the categories; Table S3: Spearman chemical correlation matrix of PCA, Table S4: Squared cosines of the variables, Table S5: Squared cosines of the variables (sectors, zones, and chemistry), Table S6: PLS correlation matrix between microphysics and chemistry, Table S7: PCA correlation matrix between microphysics and parameters related to air mass history.

Author Contributions: Conceptualization L.D. and P.R.; statistical analysis P.R.; dynamical analysis J.-L.B. and P.R.; writing—original draft preparation P.R. and L.D.; writing—review and editing, P.R., L.D., A.B., J.-L.B., A.-M.D. and M.B. All authors agreed with the submission of the manuscript. All authors have read and agreed to the published version of the manuscript.

Funding: This research received no external funding.

Acknowledgments: This work on the long-term analysis of the cloud water chemical composition was supported by the French Ministry and CNRS-INSU. Authors acknowledge additional financial support from the Observatoire de Physique du globe de Clermont-Ferrand (OPGC), from the Regional Council of Auvergne, and from the Fédération de Recherche en Environnement through the CPER Environnement founded by Région Auvergne–Rhône-Alpes, the French ministry, and the FEDER from the European community. This work was supported by CEA/CNRS contracts. The authors thank Jean-Marc Pichon and Mickaël Ribeiro for technical support and discussions.

Conflicts of Interest: The authors declare no conflict of interest.

References

1. Herckes, P.; Valsaraj, K.T.; Collett, J.L. A review of observations of organic matter in fogs and clouds: Origin, processing and fate. *Atmos. Res.* **2013**, *132–133*, 434–449. [CrossRef]
2. Herrmann, H.; Schaefer, T.; Tilgner, A.; Styler, S.A.; Weller, C.; Teich, M.; Otto, T. Tropospheric aqueous-phase chemistry: Kinetics, mechanisms, and its coupling to a changing gas phase. *Chem. Rev.* **2015**, *115*, 4259–4334. [CrossRef] [PubMed]
3. Bianco, A.; Riva, M.; Baray, J.-L.; Ribeiro, M.; Chaumerliac, N.; George, C.; Bridoux, M.; Deguillaume, L. Chemical characterization of cloud water collected at puy de Dôme by FT-ICR MS reveals the presence of SOA components. *ACS Earth Space Chem.* **2019**, *3*, 2076–2087. [CrossRef]
4. Vaïtilingom, M.; Deguillaume, L.; Vinatier, V.; Sancelme, M.; Amato, P.; Chaumerliac, N.; Delort, A.-M. Potential impact of microbial activity on the oxidant capacity and organic carbon budget in clouds. *Proc. Natl. Acad. Sci. USA* **2013**, *110*, 559–564. [CrossRef]

5. Wei, M.; Xu, C.; Chen, J.; Zhu, C.; Li, J.; Lv, G. Characteristics of bacterial community in cloud water at Mt Tai: Similarity and disparity under polluted and non-polluted cloud episodes. *Atmos. Chem. Phys.* **2017**, *17*, 5253–5270. [CrossRef]
6. Bianco, A.; Passananti, M.; Perroux, H.; Voyard, G.; Mouchel-Vallon, C.; Chaumerliac, N.; Mailhot, G.; Deguillaume, L.; Brigante, M. A better understanding of hydroxyl radical photochemical sources in cloud waters collected at the puy de Dôme station—Experimental versus modelled formation rates. *Atmos. Chem. Phys.* **2015**, *15*, 9191–9202. [CrossRef]
7. Li, J.; Wang, X.; Chen, J.; Zhu, C.; Li, W.; Li, C.; Liu, L.; Xu, C.; Wen, L.; Xue, L.; et al. Chemical composition and droplet size distribution of cloud at the summit of Mount Tai, China. *Atmos. Chem. Phys.* **2017**, *17*, 9885–9896. [CrossRef]
8. Chen, J.; Griffin, R.J.; Grini, A.; Tulet, P. Modeling secondary organic aerosol formation through cloud processing of organic compounds. *Atmos. Chem. Phys.* **2007**, *7*, 5343–5355. [CrossRef]
9. Schurman, M.I.; Boris, A.; Desyaterik, Y.; Jeffrey, L.; Collett, J. Aqueous secondary organic aerosol formation in ambient cloud water photo-oxidations. *Aerosol Air Qual. Res.* **2017**, *18*, 15–25. [CrossRef]
10. Pye, H.O.T.; Nenes, A.; Alexander, B.; Ault, A.P.; Barth, M.C.; Clegg, S.L.; Collett, J.L., Jr.; Fahey, K.M.; Hennigan, C.J.; Herrmann, H.; et al. The acidity of atmospheric particles and clouds. *Atmos. Chem. Phys.* **2020**, *20*, 4809–4888. [CrossRef]
11. Hegg, D.A.; Hobbs, P.V. Cloud water chemistry and the production of sulfates in clouds. *Atmos. Environ. 1967* **1981**, *15*, 1597–1604. [CrossRef]
12. Vong, R.J.; Hansson, H.-C.; Covert, D.S.; Charlson, R.J. Acid rain: Simultaneous observations of a natural marine background and its acidic sulfate aerosol precursor. *Geophys. Res. Lett.* **1988**, *15*, 338–341. [CrossRef]
13. Ervens, B.; Turpin, B.J.; Weber, R.J. Secondary organic aerosol formation in cloud droplets and aqueous particles (aqSOA): A review of laboratory, field and model studies. *Atmos. Chem. Phys.* **2011**, *11*, 11069–11102. [CrossRef]
14. Ervens, B.; Sorooshian, A.; Aldhaif, A.M.; Shingler, T.; Crosbie, E.; Ziemba, L.; Campuzano-Jost, P.; Jimenez, J.L.; Wisthaler, A. Is there an aerosol signature of chemical cloud processing? *Atmos. Chem. Phys.* **2018**, *18*, 16099–16119. [CrossRef]
15. Quinn, P.K.; Bates, T.S.; Coffman, D.J.; Covert, D.S. Influence of particle size and chemistry on the cloud nucleating properties of aerosols. *Atmos. Chem. Phys.* **2008**, *8*, 1029–1042. [CrossRef]
16. Bator, A.; Collett, J.L. Cloud chemistry varies with drop size. *J. Geophys. Res. Atmos.* **1997**, *102*, 28071–28078. [CrossRef]
17. Moore, K.F.; Sherman, D.E.; Reilly, J.E.; Collett, J.L. Drop size-dependent chemical composition in clouds and fogs. Part I. Observations. *Atmos. Environ.* **2004**, *38*, 1389–1402. [CrossRef]
18. Hao, L.; Romakkaniemi, S.; Kortelainen, A.; Jaatinen, A.; Portin, H.; Miettinen, P.; Komppula, M.; Leskinen, A.; Virtanen, A.; Smith, J.N.; et al. Aerosol chemical composition in cloud events by high resolution time-of-flight aerosol mass spectrometry. *Environ. Sci. Technol.* **2013**, *47*, 2645–2653. [CrossRef]
19. Van Pinxteren, D.; Fomba, K.W.; Mertes, S.; Müller, K.; Spindler, G.; Schneider, J.; Lee, T.; Collett, J.L.; Herrmann, H. Cloud water composition during HCCT-2010: Scavenging efficiencies, solute concentrations, and droplet size dependence of inorganic ions and dissolved organic carbon. *Atmos. Chem. Phys.* **2016**, *16*, 3185–3205. [CrossRef]
20. Triesch, N.; van Pinxteren, M.; Engel, A.; Herrmann, H. Concerted measurements of free amino acids at the Cape Verde Islands: High enrichments in submicron sea spray aerosol particles and cloud droplets. *Atmos. Chem. Phys. Discuss.* **2020**, 1–24. [CrossRef]
21. Van Pinxteren, D.; Plewka, A.; Hofmann, D.; Müller, K.; Kramberger, H.; Svrcina, B.; Bächmann, K.; Jaeschke, W.; Mertes, S.; Collett, J.L.; et al. Schmücke hill cap cloud and valley stations aerosol characterisation during FEBUKO (II): Organic compounds. *Atmos. Environ.* **2005**, *39*, 4305–4320. [CrossRef]
22. Wang, M.; Perroux, H.; Fleuret, J.; Bianco, A.; Bouvier, L.; Colomb, A.; Borbon, A.; Deguillaume, L. Anthropogenic and biogenic hydrophobic VOCs detected in clouds at the puy de Dôme station using Stir Bar Sorptive Extraction: Deviation from the Henry's law prediction. *Atmos. Res.* **2020**, *237*, 104844. [CrossRef]
23. Atlas, E.; Giam, C.S. Ambient concentration and precipitation scavenging of atmospheric organic pollutants. *Water. Air. Soil Pollut.* **1988**, *38*, 19–36. [CrossRef]
24. Levsen, K.; Behnert, S.; Prieß, B.; Svoboda, M.; Winkeler, H.-D.; Zietlow, J. Organic compounds in precipitation. *Chemosphere* **1990**, *21*, 1037–1061. [CrossRef]

25. Bianco, A.; Deguillaume, L.; Vaïtilingom, M.; Nicol, E.; Baray, J.-L.; Chaumerliac, N.; Bridoux, M. Molecular characterization of cloud water samples collected at the puy de Dôme (France) by Fourier Transform Ion Cyclotron Resonance Mass Spectrometry. *Environ. Sci. Technol.* **2018**, *52*, 10275–10285. [CrossRef] [PubMed]
26. Löflund, M.; Kasper-Giebl, A.; Schuster, B.; Giebl, H.; Hitzenberger, R.; Puxbaum, H. Formic, acetic, oxalic, malonic and succinic acid concentrations and their contribution to organic carbon in cloud water. *Atmos. Environ.* **2002**, *36*, 1553–1558. [CrossRef]
27. Petrenchuk, O.P.; Drozdova, V.M. On the chemical composition of cloud water. *Tellus* **1966**, *18*, 280–286. [CrossRef]
28. Fuzzi, S.; Facchini, M.C.; Decesari, S.; Matta, E.; Mircea, M. Soluble organic compounds in fog and cloud droplets: What have we learned over the past few years? *Atmos. Res.* **2002**, *64*, 89–98. [CrossRef]
29. Ghauri, B.M.; Ishaq Mirza, M.; Richter, R.; Dutkiewicz, V.A.; Rusheed, A.; Khan, A.R.; Husain, L. Composition of aerosols and cloud water at a remote mountain site (2.8 kms) in Pakistan. *Chemosphere Glob. Chang. Sci.* **2001**, *3*, 51–63. [CrossRef]
30. Guo, J.; Wang, Y.; Shen, X.; Wang, Z.; Lee, T.; Wang, X.; Li, P.; Sun, M.; Collett, J.L.; Wang, W.; et al. Characterization of cloud water chemistry at Mount Tai, China: Seasonal variation, anthropogenic impact, and cloud processing. *Atmos. Environ.* **2012**, *60*, 467–476. [CrossRef]
31. Wang, Z.; Wang, T.; Gao, R.; Xue, L.; Guo, J.; Zhou, Y.; Nie, W.; Wang, X.; Xu, P.; Gao, J.; et al. Source and variation of carbonaceous aerosols at Mount Tai, North China: Results from a semi-continuous instrument. *Atmos. Environ.* **2011**, *45*, 1655–1667. [CrossRef]
32. Aleksic, N.; Roy, K.; Sistla, G.; Dukett, J.; Houck, N.; Casson, P. Analysis of cloud and precipitation chemistry at Whiteface Mountain, NY. *Atmos. Environ.* **2009**, *43*, 2709–2716. [CrossRef]
33. Gioda, A.; Mayol-Bracero, O.L.; Scatena, F.N.; Weathers, K.C.; Mateus, V.L.; McDowell, W.H. Chemical constituents in clouds and rainwater in the Puerto Rican rainforest: Potential sources and seasonal drivers. *Atmos. Environ.* **2013**, *68*, 208–220. [CrossRef]
34. Weathers, K.C.; Likens, G.E.; Bormann, F.H.; Bicknell, S.H.; Bormann, B.T.; Daube, B.C.; Eaton, J.S.; Galloway, J.N.; Keene, W.C. Cloudwater chemistry from ten sites in North America. *Environ. Sci. Technol.* **1988**, *22*, 1018–1026. [CrossRef]
35. Crosbie, E.; Brown, M.D.; Shook, M.; Ziemba, L.; Moore, R.H.; Shingler, T.; Winstead, E.; Thornhill, K.L.; Robinson, C.; MacDonald, A.B.; et al. Development and characterization of a high-efficiency, aircraft-based axial cyclone cloud water collector. *Atmos. Meas. Tech.* **2018**, *11*, 5025–5048. [CrossRef]
36. Skarżyńska, K.; Polkowska, Ż.; Namieśnik, J. Sample handling and determination of physico-chemical parameters in rime, hoarfrost, dew, fog and cloud water samples—A review. *Pol. J. Environ. Stud.* **2005**, *15*, 185–209.
37. Wieprecht, W.; Acker, K.; Mertes, S.; Collett, J.; Jaeschke, W.; Brüggemann, E.; Möller, D.; Herrmann, H. Cloud physics and cloud water sampler comparison during FEBUKO. *Atmos. Environ.* **2005**, *39*, 4267–4277. [CrossRef]
38. Zhao, Y.; Hallar, A.G.; Mazzoleni, L.R. Atmospheric organic matter in clouds: Exact masses and molecular formula identification using ultrahigh-resolution FT-ICR mass spectrometry. *Atmos. Chem. Phys.* **2013**, *13*, 12343–12362. [CrossRef]
39. Baray, J.-L.; Deguillaume, L.; Colomb, A.; Sellegri, K.; Freney, E.; Rose, C.; Baelen, J.V.; Pichon, J.-M.; Picard, D.; Fréville, P.; et al. Cézeaux-Aulnat-Opme-Puy De Dôme: A multi-site for the long term survey of the tropospheric composition and climate change. *Atmos. Meas. Tech.* **2020**, *13*, 3413–3445. [CrossRef]
40. Acker, K. Mt. Brocken, a site for a cloud chemistry measurement programme in Central Europe. *Water Air Soil Pollut.* **1995**, *4*, 6.
41. Takeuchi, M.; Okochi, H.; Igawa, M. Characteristics of water-soluble components of atmospheric aerosols in Yokohama and Mt. Oyama, Japan from 1990 to 2001. *Atmos. Environ.* **2004**, *38*, 4701–4708. [CrossRef]
42. Science, G. Experimental evidence of the feeding of the free troposphere with aerosol particles from the mixing layer. *Aerosol Air Qual. Res.* **2016**, *16*, 702–716. [CrossRef]
43. Putaud, J.-P.; Raes, F.; Van Dingenen, R.; Brüggemann, E.; Facchini, M.-C.; Decesari, S.; Fuzzi, S.; Gehrig, R.; Hüglin, C.; Laj, P.; et al. A European aerosol phenomenology—2: Chemical characteristics of particulate matter at kerbside, urban, rural and background sites in Europe. *Atmos. Environ.* **2004**, *38*, 2579–2595. [CrossRef]

44. Clain, G.; Baray, J.-L.; Delmas, R.; Keckhut, P.; Cammas, J.-P. A lagrangian approach to analyse the tropospheric ozone climatology in the tropics: Climatology of stratosphere–troposphere exchange at Reunion Island. *Atmos. Environ.* **2010**, *44*, 968–975. [CrossRef]
45. Hondula, D.M.; Sitka, L.; Davis, R.E.; Knight, D.B.; Gawtry, S.D.; Deaton, M.L.; Lee, T.R.; Normile, C.P.; Stenger, P.J. A back-trajectory and air mass climatology for the Northern Shenandoah Valley, USA. *Int. J. Climatol.* **2010**, *30*, 569–581. [CrossRef]
46. Deguillaume, L.; Charbouillot, T.; Joly, M.; Vaïtilingom, M.; Parazols, M.; Marinoni, A.; Amato, P.; Delort, A.-M.; Vinatier, V.; Flossmann, A.; et al. Classification of clouds sampled at the puy de Dôme (France) based on 10 yr of monitoring of their physicochemical properties. *Atmos. Chem. Phys.* **2014**, *14*, 1485–1506. [CrossRef]
47. Deininger, C.K.; Saxena, V.K. A validation of back trajectories of air masses by principal component analysis of ion concentrations in cloud water. *Atmos. Environ.* **1997**, *31*, 295–300. [CrossRef]
48. Kim, M.-G.; Lee, B.-K.; Kim, H.-J. Cloud/fog water chemistry at a high elevation site in South Korea. *J. Atmos. Chem.* **2006**, *55*, 13–29. [CrossRef]
49. Yue, Y.; Niu, S.; Zhao, L.; Zhang, Y.; Xu, F. The influences of macro- and microphysical characteristics of sea-fog on fog-water chemical composition. *Adv. Atmos. Sci.* **2014**, *31*, 624–636. [CrossRef]
50. Baray, J.-L.; Bah, A.; Cacault, P.; Sellegri, K.; Pichon, J.-M.; Deguillaume, L.; Montoux, N.; Noel, V.; Seze, G.; Gabarrot, F.; et al. Cloud occurrence frequency at puy de Dôme (France) deduced from an automatic camera image analysis: Method, validation, and comparisons with larger scale parameters. *Atmosphere* **2019**, *10*, 808. [CrossRef]
51. Brantner, B.; Fierlinger, H.; Puxbaum, H. Cloudwater chemistry in the subcooled droplet regime at Mount Sonnblick (3106 M A.S.L., Salzburg, Austria). *Water Air Soil Pollut.* **1993**, *74*, 362–384.
52. Lebedev, A.T.; Polyakova, O.V.; Mazur, D.M.; Artaev, V.B.; Canet, I.; Lallement, A.; Vaïtilingom, M.; Deguillaume, L.; Delort, A.-M. Detection of semi-volatile compounds in cloud waters by GC×GC-TOF-MS. Evidence of phenols and phthalates as priority pollutants. *Environ. Pollut.* **2018**, *241*, 616–625. [CrossRef] [PubMed]
53. Wirgot, N.; Vinatier, V.; Deguillaume, L.; Sancelme, M.; Delort, A.-M. H_2O_2 modulates the energetic metabolism of the cloud microbiome. *Atmos. Chem. Phys.* **2017**, *17*, 14841–14851. [CrossRef]
54. Renard, P.; Canet, I.; Sancelme, M.; Wirgot, N.; Deguillaume, L.; Delort, A.-M. Screening of cloud microorganisms isolated at the Puy de Dôme (France) station for the production of biosurfactants. *Atmos. Chem. Phys.* **2016**, *16*, 12347–12358. [CrossRef]
55. Lallement, A.; Besaury, L.; Tixier, E.; Sancelme, M.; Amato, P.; Vinatier, V.; Canet, I.; Polyakova, O.V.; Artaev, V.B.; Lebedev, A.T.; et al. Potential for phenol biodegradation in cloud waters. *Biogeosciences* **2018**, *15*, 5733–5744. [CrossRef]
56. Stookey, L.L. Ferrozine—A new spectrophotometric reagent for iron. *Anal. Chem.* **1970**, *42*, 779–781. [CrossRef]
57. Amato, P.; Joly, M.; Besaury, L.; Oudart, A.; Taib, N.; Moné, A.I.; Deguillaume, L.; Delort, A.-M.; Debroas, D. Active microorganisms thrive among extremely diverse communities in cloud water. *PLoS ONE* **2017**, *12*, e0182869. [CrossRef]
58. Amato, P.; Besaury, L.; Joly, M.; Penaud, B.; Deguillaume, L.; Delort, A.-M. Metatranscriptomic exploration of microbial functioning in clouds. *Sci. Rep.* **2019**, *9*, 4383. [CrossRef]
59. Bianco, A.; Voyard, G.; Deguillaume, L.; Mailhot, G.; Brigante, M. Improving the characterization of dissolved organic carbon in cloud water: Amino acids and their impact on the oxidant capacity. *Sci. Rep.* **2016**, *6*, 37420. [CrossRef]
60. Bianco, A.; Vaïtilingom, M.; Bridoux, M.; Chaumerliac, N.; Pichon, J.-M.; Piro, J.-L.; Deguillaume, L. Trace metals in cloud water sampled at the puy de Dôme station. *Atmosphere* **2017**, *8*, 225. [CrossRef]
61. Bianco, A.; Deguillaume, L.; Chaumerliac, N.; Vaïtilingom, M.; Wang, M.; Delort, A.-M.; Bridoux, M.C. Effect of endogenous microbiota on the molecular composition of cloud water: A study by Fourier-transform ion cyclotron resonance mass spectrometry (FT-ICR MS). *Sci. Rep.* **2019**, *9*, 7663. [CrossRef] [PubMed]
62. Martinsson, J.; Monteil, G.; Sporre, M.K.; Kaldal Hansen, A.M.; Kristensson, A.; Eriksson Stenström, K.; Swietlicki, E.; Glasius, M. Exploring sources of biogenic secondary organic aerosol compounds using chemical analysis and the FLEXPART model. *Atmos. Chem. Phys.* **2017**, *17*, 11025–11040. [CrossRef]
63. Van Pinxteren, D.; Brüggemann, E.; Gnauk, T.; Müller, K.; Thiel, C.; Herrmann, H. A GIS based approach to back trajectory analysis for the source apportionment of aerosol constituents and its first application. *J. Atmos. Chem.* **2010**, *67*, 1–28. [CrossRef]

64. Hoffmann, L.; Günther, G.; Li, D.; Stein, O.; Wu, X.; Griessbach, S.; Heng, Y.; Konopka, P.; Müller, R.; Vogel, B.; et al. From ERA-Interim to ERA5: The considerable impact of ECMWF's next-generation reanalysis on Lagrangian transport simulations. *Atmos. Chem. Phys.* **2019**, *19*, 3097–3124. [CrossRef]
65. Bezdek, A.; Sebera, J. MATLAB script for visualizing geoid height and other elevation data on rotating 3D globe. *Comput. Geosci.* **2013**, *56*, 127–130. [CrossRef]
66. Ilin, A.; Raiko, T. Practical Approaches to principal component analysis in the presence of missing values. *J. Mach. Learn. Res.* **2010**, *11*, 1957–2000.
67. Addinsoft Addinsoft. XLSTAT Statistical and Data Analysis Solution. New York, NY, USA, 2020. Available online: https://www.xlstat.com (accessed on 23 May 2020).
68. Błaś, M.; Polkowska, Ż.; Sobik, M.; Klimaszewska, K.; Nowiński, K.; Namieśnik, J. Fog water chemical composition in different geographic regions of Poland. *Atmos. Res.* **2010**, *95*, 455–469. [CrossRef]
69. Collett, J.L.; Bator, A.; Sherman, D.E.; Moore, K.F.; Hoag, K.J.; Demoz, B.B.; Rao, X.; Reilly, J.E. The chemical composition of fogs and intercepted clouds in the United States. *Atmos. Res.* **2002**, *64*, 29–40. [CrossRef]
70. Nieberding, F.; Breuer, B.; Braeckevelt, E.; Klemm, O.; Song, Q.; Zhang, Y. Fog Water Chemical Composition on Ailaoshan Mountain, Yunnan Province, SW China. *Aerosol Air Qual. Res.* **2017**, *18*, 37–48. [CrossRef]
71. Wang, Y.; Zhang, J.; Marcotte, A.R.; Karl, M.; Dye, C.; Herckes, P. Fog chemistry at three sites in Norway. *Atmos. Res.* **2015**, *151*, 72–81. [CrossRef]
72. Xu, X.; Chen, J.; Zhu, C.; Li, J.; Sui, X.; Liu, L.; Sun, J. Fog composition along the Yangtze River basin: Detecting emission sources of pollutants in fog water. *J. Environ. Sci.* **2018**, *71*, 2–12. [CrossRef] [PubMed]
73. Möller, D.; Acker, K.; Wieprecht, W. A relationship between liquid water content and chemical composition in clouds. *Atmos. Res.* **1996**, *41*, 321–335. [CrossRef]
74. Collett, J.L.; Daube, B.C.; Gunz, D.; Hoffmann, M.R. Intensive studies of Sierra Nevada cloudwater chemistry and its relationship to precursor aerosol and gas concentrations. *Atmos. Environ. Part Gen. Top.* **1990**, *24*, 1741–1757. [CrossRef]
75. Watanabe, K.; Ishizaka, Y.; Takenaka, C. Chemical characteristics of cloud water over the Japan Sea and the Northwestern Pacific Ocean near the central part of Japan: Airborne measurements. *Atmos. Environ.* **2001**, *35*, 645–655. [CrossRef]
76. Bourcier, L.; Sellegri, K.; Chausse, P.; Pichon, J.M.; Laj, P. Seasonal variation of water-soluble inorganic components in aerosol size-segregated at the puy de Dôme station (1465 m a.s.l.), France. *J. Atmos. Chem.* **2012**, *69*, 47–66. [CrossRef]
77. Fu, P.; Kawamura, K.; Kobayashi, M.; Simoneit, B.R.T. Seasonal variations of sugars in atmospheric particulate matter from Gosan, Jeju Island: Significant contributions of airborne pollen and Asian dust in spring. *Atmos. Environ.* **2012**, *55*, 234–239. [CrossRef]
78. Shapiro, J.B.; Simpson, H.J.; Griffin, K.L.; Schuster, W.S.F. Precipitation chloride at West Point, NY: Seasonal patterns and possible contributions from non-seawater sources. *Atmos. Environ.* **2007**, *41*, 2240–2254. [CrossRef]
79. Kundu, S.; Kawamura, K.; Andreae, T.W.; Hoffer, A.; Andreae, M.O. Diurnal variation in the water-soluble inorganic ions, organic carbon and isotopic compositions of total carbon and nitrogen in biomass burning aerosols from the LBA-SMOCC campaign in Rondônia, Brazil. *J. Aerosol Sci.* **2010**, *41*, 118–133. [CrossRef]
80. Elbert, W.; Krämer, M.; Andreae, M.O. Reply to discussion on "Control of solute concentrations in cloud and fog water by liquid water content". *Atmos. Environ.* **2002**, *11*, 1909–1910. [CrossRef]
81. Fahey, K.M.; Pandis, S.N.; Collett, J.L.; Herckes, P. The influence of size-dependent droplet composition on pollutant processing by fogs. *Atmos. Environ.* **2005**, *39*, 4561–4574. [CrossRef]
82. Reilly, J.E.; Rattigan, O.V.; Moore, K.F.; Judd, C.; Eli Sherman, D.; Dutkiewicz, V.A.; Kreidenweis, S.M.; Husain, L.; Collett, J.L. Drop size-dependent S(IV) oxidation in chemically heterogeneous radiation fogs. *Atmos. Environ.* **2001**, *35*, 5717–5728. [CrossRef]
83. Herrmann, H.; Exner, M.; Zellner, R. Reactivity trends in reactions of the nitrate radical (NO_3) with inorganic and organic cloudwater constituents. *Geochim. Cosmochim. Acta* **1994**, *58*, 3239–3244. [CrossRef]
84. Heuvel, A.P.V.D.; Mason, B.J. The formation of ammonium sulphate in water droplets exposed to gaseous sulphur dioxide and ammonia. *Q. J. R. Meteorol. Soc.* **1963**, *89*, 271–275. [CrossRef]
85. Hung, H.-M.; Hoffmann, M.R. Oxidation of Gas-Phase SO_2 on the Surfaces of Acidic Microdroplets: Implications for Sulfate and Sulfate Radical Anion Formation in the Atmospheric Liquid Phase. *Environ. Sci. Technol.* **2015**, *49*, 13768–13776. [CrossRef] [PubMed]

86. Kwong, K.C.; Chim, M.M.; Davies, J.F.; Wilson, K.R.; Chan, M.N. Importance of sulfate radical anion formation and chemistry in heterogeneous OH oxidation of sodium methyl sulfate, the smallest organosulfate. *Atmos. Chem. Phys. Online* **2018**, *18*. [CrossRef]
87. Rose, C.; Chaumerliac, N.; Deguillaume, L.; Perroux, H.; Mouchel-Vallon, C.; Leriche, M.; Patryl, L.; Armand, P. Modeling the partitioning of organic chemical species in cloud phases with CLEPS (1.1). *Atmos. Chem. Phys.* **2018**, *18*, 2225–2242. [CrossRef]

atmosphere

MDPI

Article

Physico-Chemical Characterization of an Urban Rainwater (Zagreb, Croatia)

Palma Orlović-Leko [1], Kristijan Vidović [2,*], Irena Ciglenečki [1], Dario Omanović [1], Mathieu Dutour Sikirić [1] and Ivan Šimunić [3]

[1] Division for Marine and Environmental Research, Ruđer Bošković Institute, Bijenička 54, 10000 Zagreb, Croatia; palmaorlovic@gmail.com (P.O.-L.); irena@irb.hr (I.C.); omanovic@irb.hr (D.O.); mathieu.andre.dutour-sikiric@irb.hr (M.D.S.)

[2] National Institute of Chemistry, Hajdrihova 19, SI-1000 Ljubljana, Slovenia

[3] Faculty of Agriculture, Department of Soil Amelioration, University of Zagreb, Svetošimunska cesta 25, 10000 Zagreb, Croatia; simunic@agr.hr

* Correspondence: kristijan.vidovic@ki.si

Received: 20 December 2019; Accepted: 22 January 2020; Published: 28 January 2020

Abstract: The characterization of organic matter (OM) and trace elements (TEs) was conducted in bulk precipitation samples collected in an urban area of Croatia (Zagreb center), from January 2009 to October 2011 (N = 31). Characterization of OM was performed by measurement of dissolved (DOC) and particulate forms of organic carbon (POC), as well as surface-active substances, copper complexing capacity and reduced sulfur species which were determined electrochemically. Concentrations of TEs (Al, Fe, Zn, Cr, Ba, Mn, Cu, Sr, Ti, Pb, V, Ni, Rb, Sb, As, Sn, Se, Co, Cd, Mo) were analyzed by HR ICP-MS. The most important outcome of this study is confirmed relatively low concentrations of DOC (0.69–4.86 mgC L^{-1}) and TEs. Daily fluxes of Zn, Pb, Ni, As, and Cd were two to three times lower than that of those reported for an urban industrial site in Europe. Additionally, this study shows that the value and reactivity of rainwater DOC has not considerably changed if compared with the results from the 1998–1999 study (0.78–4.39 mgC L^{-1}). High traffic density and thermal power plants are assumed to be the main local sources of pollutants in Zagreb. Using Al as a reference element, it was found that Cd, Sb, and Cu were associated with anthropogenic sources. The solubility of Sr, Zn, Cd, Ni, Cr, As, and Rb was higher than 70%. Another critical characteristic of precipitation composition is episodic variation in POC concentration because of Saharan dust transport. Obtained data can be valuable for environmental quality assessment, as well as for insight into atmospheric deposition processes.

Keywords: organic carbon; surface-active substances; trace elements; reduced sulfur species; long-range transport

1. Introduction

Many atmospheric organic and inorganic compounds have a tendency toward dissolution in atmospheric water (cloud droplets and rainwater) [1]. Rainwater is very sensitive to changes in the chemistry of the atmosphere [2]. The physico-chemical properties of the precipitation are influenced by local sources (natural and anthropogenic) [1] as well as long-range transport of dust and aerosols [3,4]. It was found that 40% of the dissolved organic carbon (DOC) in the rainwater is resistant to bacterial degradation and may be transported by long-distance in the atmosphere before being removed by rain [5]. Previous studies on the rainwater chemistry in Croatia have shown that the rainwater composition is profoundly affected by the transport of the eolian dust from North Africa (very often in spring and summer) as well as polluted aerosols from the other Europe countries [6,7]. Atmospheric transport and deposition is an important pathway of various substances, including organic compounds and trace metals to the surface environment [5,8–13]. Removal of substances from the atmosphere by precipitation affects chemistry, biogeochemistry, and sedimentation in the aquatic environment [11,14]. These processes have a decisive role in global changes and in environmental protection [5,8–11].

Dissolved organic matter (DOM), measured as dissolved organic carbon (DOC), is an important component of the rainwater [5]. Removal of the atmospheric DOC by rainwater, including non-oxidized organic compounds, decreases the generation of carbon dioxide [15]. This process is crucial for the evaluation of the global carbon cycle [5,15]. The climate change in recent decades has led to increased interest in the study of DOC in atmospheric precipitation [16–20] and surface waters [14] (and references therein).

Surface-active substances (SAS) make up a portion of rainwater DOC [21–23] (and references therein) and the water-soluble organic carbon (WSOC) fraction of atmospheric aerosols [24] (and references therein). These compounds are the most reactive part of DOC [25] because they tend to adsorb on the natural phase boundaries (air-water, water-sediment/suspended particles) and, in this way, significantly affect the transfer of mass and energy. In our previous work, we have shown that some metal ions can accumulate in adsorbed layers of organic molecules [23]. This mechanism could be important for the geochemical cycle of metals.

Film-forming compounds can influence the microphysical processes of humid aerosol particles and atmospheric droplets [24,26]. Because of these properties, atmospheric SAS plays an essential role in various atmospheric processes that reflect climate change [24] (and references therein). Physico-chemical properties of OM are often more important than the amount present [22]. Atmospheric humic-like substances, HULIS were found to be an important class of SAS in aerosol particles [24,27,28] (and references therein). Rainwater SAS are compositionally different from those found in surface waters [29]. Atmospheric HULIS decrease the surface tension more efficiently than the terrestrial and aquatic humic substances because of the different composition [27]. Some studies also suggested that atmospheric SAS could be "biosurfactants" of microbial origin [24]. Quantification and characterization of the SAS in the atmospheric deposition are important for a better understanding of its contribution to organic carbon (OC) cycling.

Complexation of OC with metal ions has a significant impact on mobility, solubility, and bioavailability of metals in precipitation and further in the aquatic environment [30–36].

The organically complexed metals such as Cu, Co, Ni, Pb, Cd, and Zn have been observed in urban rainwater [32] (and references therein). Metal complexing capacity (MeCC) refers to the amount of OM present that can specifically bind metal ions, and it is often used as a parameter for qualitative and quantitative characterization of the OM in natural waters [32–35]. These investigations are usually focused on the complexation of Cu ligands [32–35] because the Cu ions have a strong affinity for naturally occurring organic ligands [32,33]. In the urban atmosphere, Cu is derived from fossil fuel combustion, exhaust emissions, and industrial processes [37] (and references therein), although the significant atmospheric Cu in western Europe comes from brake wear [38].

In the atmosphere, sulfur is mainly emitted in its reduced form [39]. Carbon disulfide (CS_2), dimethyl sulfide (CH_3SCH_3, DMS), and hydrogen sulfide (H_2S) are the most frequently reduced sulfur species (RSS), which are found in the atmosphere [40]. In Orlović-Leko et al. [41], inorganic or organic RSS has been confirmed at nano-levels in rainwater samples by electrochemical methods. The chemistry of RSS is crucial for many environmental problems, such as acid rain and climate change [39]. While sulfur species, together with SAS, could play a crucial role in speciation and solubility of trace metals in aquatic systems [42].

In an urban environment, metals are emitted primarily by anthropogenic sources, which include traffic, industry, corrosion of construction materials, and waste incineration [43].

An important ecological property of metals is their persistence and ability to accumulate in ecosystems [43]. Measurements of TEs concentrations in the rainwater could be critical to understanding of their geochemical cycling [44–47]. The investigation of the soluble fraction of metal in atmospheric deposition is a priority in many studies [36,47–49] because the solubility of TEs affects their bioavailability. The solubility of TEs depends on the pH values of rainwater and the size of particles [48]. The particles of anthropogenic origin are highly soluble [46] (and references therein).

The aim of this study is the characterization of OM (through measurement of DOC, particulate organic carbon POC, SAS, RSS) and TEs in precipitation of an urban environment, in Zagreb city, Croatia. In our previous studies, conducted in 1998–1999 and 2003–2007, spatial and temporal variability of DOC concentrations in bulk precipitation of urban and coastal areas in Croatia have been discussed [21,22]. However, in accordance with the literature that emphasized a lack of data on DOC measurements in rainwater in central and eastern Europe [17], as well as trends based on the long-term observations of DOC in rainwater [8] (and references therein), in this paper, we focus on studying the possible changes that may have occurred in the concentration and reactivity of DOC between two sampling periods (1998–1999 and 2009–2011), in line with socio-economic and global changes. Namely, Croatia ratified the Kyoto Protocol (April 2007), thereby committing to reducing emissions of greenhouse gases in the period 2008–2012 compared to the baseline emissions in nineties (1990). Also, as a prerequisite for achievements of EU Directive 2006/32/EC on energy efficiency and energy services, Croatia prepared a program "Adjustment and upgrades of energy development strategies," which foresees adaptation of the power plant in Zagreb to use renewable energy. In the studied sampling period the Zagreb thermal power plant was operated on fuel oil, which may affect DOC and TE concentrations. Therefore, this study on the rainwater OM and TE is unique for Zagreb, and Croatia, and could serve as a baseline for future investigations in the area.

Data treatment for TEs included the calculation of their daily fluxes and percentage of solubility. In order to study the possible sources of TEs, correlation analyses, and enrichment factors (EFs) were considered.

2. Experiments

2.1. Area Description

The sampling site (Figure 1) was located in the center of Zagreb (45°48′24″ N and 15°57′50″ E), on the roof (20 m above ground level) of the Faculty of Mining, Geology, and Petroleum Engineering building. The site is not surrounded by rows of buildings, and therefore, airflow was not channeled, obstructed, or restricted. It is close to the parking lot and two main roads characterized by heavy traffic all day long. The thermal power station (burns gas and oil fuel) is located approximately 2 km from our sampling location. However, it is important to note that there are no large industries in Zagreb and that the topography of the sampling site did not change over the course of the sampling period. The primary local pollution sources are traffic and two thermal power stations. Gas is a standard fuel for domestic heating.

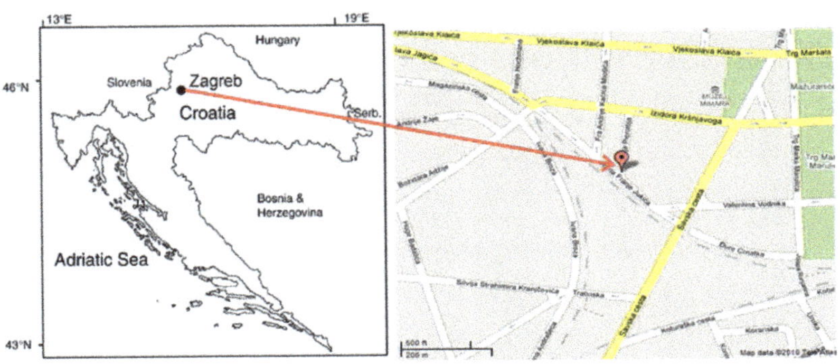

Figure 1. Position of the sampling site (adapted from Google Maps).

Zagreb has a continental precipitation regime, with its maximum in the warm months of the year and a secondary maximum in autumn. The analysis of annual precipitation amounts from 1961 to 1990 showed that in Croatia, the annual precipitation amount in 2009 (794.8 mm) and 2011 (520.8 mm) was below the average, and in 2010 (1155.1 mm) it was above the average [50–52]. A comparison of monthly precipitation in 2009/2010 in the Zagreb area is presented in Figure 2 [50,51]. During the study period, daily precipitation ranged from 0.3 to 28.8 mm (Croatian Meteorological and Hydrological Service Data, DHZM).

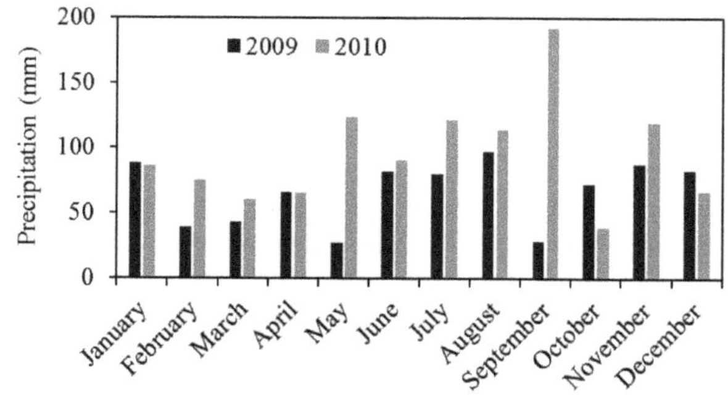

Figure 2. Monthly precipitation 2009/2010 at Zagreb area [50,51].

2.2. *Sampling Method and Sample Treatment Prior Analyses*

Daily, bulk precipitation (27 samples) were sampled from January 2009 to May 2010. In 2011, only four samples were collected, each in different seasons. The sampling was performed using two samplers consisting of a glass funnel (25 cm diameter) and a glass bottle (2.5 L), which were installed 1.0 m above the ground for bulk rainwater sampling. Both components of the sampler were prepared according to the standard cleaning procedure. They were washed with chrome-sulphuric acid (in-house), rinsed by HNO_3 (p.a. Kemika, Zagreb, Croatia), and several times with Milli-Q water (Milli-Q, 18.2 MW, total organic carbon (TOC) < 3 ppb) before being set up at the sampling location. The collected precipitation volumes ranged from 0.1 to 2.2 L (collected samples with a volume of less than 0.1 L were not considered).

The bulk sampler (bottle/funnel) collected dry deposition of gases and particles. However, it was found that approximately 80% of carbonaceous aerosols are removed by wet deposition (via precipitation) [53]. Furthermore, according to the European Committee for Standardization (CEN), the bulk sampler can also be used for measurements of atmospheric deposition of metals in industrial and urban sites, mainly in the case of daily samples [6,47] (and reference therein).

Volumes and the pH values of samples were measured just after collection. For measurements of TEs in the filtered fraction, the samples were filtered by use of the Millipore vacuum filtration system through a 0.45 μm cellulose-nitrate membrane filter (Sartorius). For DOC measurement, the filtered fraction was obtained by the use of the same filtration system with Whatman glass fiber filters, pore size 0.7 μm (Whatman, Grade GF/F, d = 47 mm). After preservation with the $HgCl_2$ solution, it was stored in a cold dark place until analysis. Particulate organic carbon (POC), defined as organic matter larger than 0.7 μm, remained on the filter. Parameters, SAS, CuCC, and RSS, were measured in the original samples within two days. Prior to analysis, the samples were stored at 4 °C.

2.3. *Chemical Analyses*

2.3.1. Organic Matter (DOC, POC Measurements)

DOC and POC content was determined by the high-temperature catalytic oxidation (HTCO) method at a TOC-VCPH instrument (Shimadzu, Japan). The DOC and POC concentrations of each sample were calculated as an average of three replicates. The limits of quantification (LOQ) are 0.228 mg/L for DOC and 7.11 μg/L for POC. The precision of DOC and POC measurements, given as the relative standard deviation (RSD) was based on the analysis of selected samples and the reference materials; RSD never exceeded 6% and 5%, respectively. Both measurement procedures were validated through international intercalibrations as a prerequisite for accreditation achieved in 2017 (HRN EN ISO/IEC 17025:2007).

2.3.2. Surface Activity of Rainwater DOC

The surface activity of DOC was determined by the electrochemical method of alternating current voltammetry with out-of-phase mode by using a μ-Autolab (Electrochemical Instrument Eco Chemie, Metrohm Autolab B.V., Utrecht, The Netherlands) Potentiostat connected with 663 VA Stand Metrohm mercury electrode [54]. The concentration of surface-active substances (SAS) was expressed as equivalent in mg/L to a model substance, the nonionic surfactant polyoxy ethylene-*t*-octylphenol (Triton-X-100). The detection limit of SAS determination was 0.01 mg L^{-1} equivalent of T-X-100, with LOQ of 0.03 mg L^{-1}. The method enables a rough characterization of SAS based on its adsorption behavior, i.e., hydrophobic—hydrophilic interactions, as seen by electrochemical measurements at the Hg electrode in the water solution [54].

2.3.3. CuCC Measurements

CuCC was measured by using differential pulse anodic stripping voltammetry (DPASV) by direct titration of the sample with copper ions [34]. The value of CuCC, as well as the corresponding stability constant, was calculated by applying the linear transformation plot [34] (and references therein). The RSD of the mean value calculated for five independent measurements was below 10%.

2.3.4. Reduced Sulfur Species

The RSS measurements were performed by cathodic stripping linear sweep voltammetry with a μ-Autolab (Electrochemical Instrument Eco Chemie, Utrecht, The Netherlands) electrochemical analyzer connected to a 663 VA Stand Metrohm mercury electrode, as previously described [55]. Quantification for RSS was done based on sulfide calibration. The precision obtained with 120 s accumulation time is 0.5 nM (1 σ), giving a detection limit (3 σ) of about 1.5 nM, with relative 1σ uncertainty better than ±10%.

2.3.5. Trace Elements

Multielemental analysis was performed by Element 2, High Resolution Inductively Coupled Plasma Mass Spectrometer (HR ICPMS, Thermo, Bremen, Germany). For the analyses, the samples were prepared in pre-cleaned polypropylene tubes by adding 100 μL of concentrated HNO_3 and 50 μL of In (115) as internal standard (0.1 mg L^{-1}) into 5 mL of the sample aliquot. The concentrations of the elements were determined using the external calibration plots. Quality control (QC) of HR ICP-MS measurements was checked by the determination of element concentration in "River Water Reference Material for Trace Metals" (SLRS-4, National Research Council Canada). Measured concentrations of all elements in the reference material were within the range of provided certified uncertainties (Table S1 in Supplementary Materials). The procedural blanks of MQ water were far below the level of concentrations measured in samples.

2.3.6. pH Measurements

A digital pH meter (Model Metrohm 691) was used for the determination of the pH parameter values. Calibration was carried out before measurement using standard buffer solutions of pH 4.00 and 7.00.

2.3.7. Correlation Analysis

For the purpose of finding the relation between investigated parameters, a correlation analysis was carried out. For this, Real Statistics was used, which is an add-in tool for Excel (Microsoft office). Before correlation analysis, each variable (concentration of metals, pH, rainfall, the solubility of metals, and concentration of DOC) was tested for normality by use of Shapiro-Wilk's test (for which Real Statistics add-in was also used). In the case where the variables did not show normality or linearity, a Spearman's rank correlation was performed, where in the case of normal distributed variables with linear relationships between them, a Pearson correlation analysis was used.

3. Results and Discussion

3.1. *Organic Carbon*

Temporal variability of concentrations of DOC and POC in the rainwater samples from January 2009 through October 2011 (N = 31) in Zagreb city center are shown in Figure 3. The daily sample concentrations of DOC ranged from 0.69 to 4.86 mgC L^{-1}, with an average value of 1.82 ± 1.02 mgC L^{-1}. The DOC content found in this study was much lower than the concentrations reported for other urban and polluted areas [17–19,56]. For example, in the industrialized and urbanized region of Poznań (Poland), the average DOC concentration (5.10 mgC L^{-1}) was about three times higher than values in Zagreb [17]; in Mexico City DOC (up to 19.5 mgC L^{-1}) is five times higher [19], while in the southeastern Tibetan Plateau, DOC ranges from 0.74 to 1.31 mgC L^{-1} [20]. This difference can be partly attributed to the various origins of DOC in precipitation. In Zagreb, rainwater DOC concentrations were mainly associated with traffic (both diesel and gasoline exhausts) and thermal power stations (combustion of natural gas and/or fuel oil).

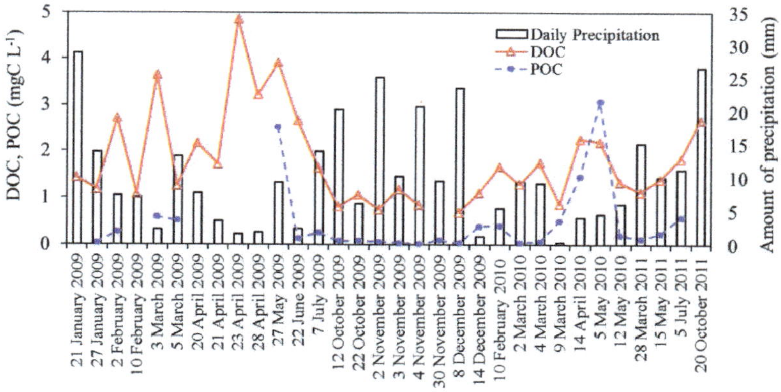

Figure 3. Temporal variability of dissolved organic carbon (DOC), particulate forms of organic carbon (POC) concentrations in rainwater in Zagreb, and daily amount of precipitation during the sampling period (2009–2011).

The air mass sources and rainfall amount can affect rainwater DOC concentrations [18,20]. At our location, a weak and significant negative correlation (r = −0.39: $p < 0.05$) between concentrations of DOC and rainfall amount was found, similar to what has been observed in our previous study [22]. According to Kieber et al. [15], this result may reflect the possibility that some components of DOC are formed in the atmospheric aqueous phase continuously during rain events. However, this lack of correlation also may be due to the photolysis of DOC [20].

No significant correlation ($p > 0.05$) was observed between DOC concentrations and pH values (Table S2 in Supplementary Materials). Results suggest that organic acids contributed a relatively small percentage to the acidity in the Zagreb precipitation [15].

The DOC concentrations from this study are not significantly different from the data of our previous two studies conducted in the Zagreb area from 1998 to 1999 and 2003 to 2007 (0.78–4.39 and 0.67–4.03 mgC L^{-1}) [21,22]. Additionally, a very similar average concentration of DOC (1.85 ± 0.81 mgC L^{-1}) was calculated for the sampling period from 1998 to 1999 at the same location (center of Zagreb) [21]. In contrast, a rainwater study conducted in Wilmington, NC, USA, observed an approximately 50% reduction of DOC within a decade (between 1996–1997 and 2008) [2]. The authors attribute these results to changes in air quality regulation and advances in emission technology.

In order to study the possible sources of DOC, we applied correlation analysis between DOC and TEs concentrations measured in the same samples (see the section below), Table S3 in Supplementary Materials. The significant ($p < 0.05$) and moderate correlation values (Spearman's coefficient, $0.50 < r < 0.70$) are found for anthropogenic metals (tracers of vehicular emissions): Cu, Fe, Ni, Pb, Sb, and Zn [37,38]. Among the fossil fuel combustion-generated elements [37], DOC significantly correlated ($p \leq 0.05$; $r > 0.60$) with As and Co and especially correlated highly with Se and Mo ($r = 78$). There is a good relationship between DOC and soil elements: Fe, Al, Mn, and Sr ($p < 0.005$; $0.70 < r < 0.76$). A significant link between DOC and metals point to their possible association and to the similarity of their sources. It is known that DOC can act as a chelating reagent and thus influences the solubility of the metal in precipitation [31] and in the aquatic environment [35].

The concentrations of DOC obtained in this study for continental rainwater were compared with those of surface freshwater systems in Croatia, Table 1.

Table 1. Comparison of DOC concentrations between rainwater and natural surface waters in Croatia.

Aquatic System (Croatia)	DOC (mgC L^{-1})	References
Lake Brljan (Krka River)	0.505–1.530	[57]
River Sava (Zagreb)	1.14–2.31 [a]	
River Lonja	4.62–7.75	[58,59]
River Drava (Osijek)	4.91	[60]
River Dunav (Batina)	4.21	[60]
Rainwater (Zagreb)	0.69–4.86 [b]	

[a] Data 2016; [b] Data 2009–2011.

As can be seen, the concentrations of DOC in rainwater are in the same concentration range as measured in the selected freshwater systems, except for Lake Brljan (Krka River), which has relatively low DOC values. It is important to note that the Krka River is characterized by extremely low natural concentrations of trace elements also [61]. Therefore, it can be assumed that this ecosystem could be affected by the input of atmospheric organic and inorganic compounds. In the study conducted in Pennsylvania (USA) [9], the average annual volume-weighted concentration of DOC was 0.71 mgC L^{-1}, and annual wet deposition fluxes of DOC were in the range from 3 to 13 kg C ha^{-1} $year^{-1}$, with an average value of 8 kg C ha^{-1} $year^{-1}$. This work underlined the significance of the atmospheric deposition of reactive carbon species to watersheds. Bao et al. [10] reported that precipitation inputs play a significant role in supplying the fraction of biolabile DOC (BDOC) to the ocean. They estimated the depositional flux of rainwater DOC into Xiamen Bay (southeastern China) of 2.1 Gg C $year^{-1}$; approximately 35% of that was BDOC flux.

Particulate organic carbon is very often used to quantify OM in natural waters [8]. In the present study, concentration of POC (N = 25) in rainwater ranged from 0.03 to 3.09 mgC L^{-1} (average 0.48 ± 0.77), Figure 3. For the majority of samples (N = 18), concentrations of POC were less than 0.5 mgC L^{-1}. According to the literature [62], rainwater usually contains between 0.2–0.5 mgC L^{-1} of POC.

Higher POC values were observed in samples from 27 May 2009 (POC = 2.52 mgC L^{-1}; pH = 6.74), 14 April 2010 (POC = 1.46 mgC L^{-1}; pH = 5.4) and 5 May 2010 (POC = 3.09 mgC L^{-1}; pH = 7.25). The last sample, with the maximum POC value (>DOC = 2.2 mgC L^{-1}), was sampled during the Saharan dust event as shown in Figure 4 for Zagreb area from 24 April to 5 May in 2010.

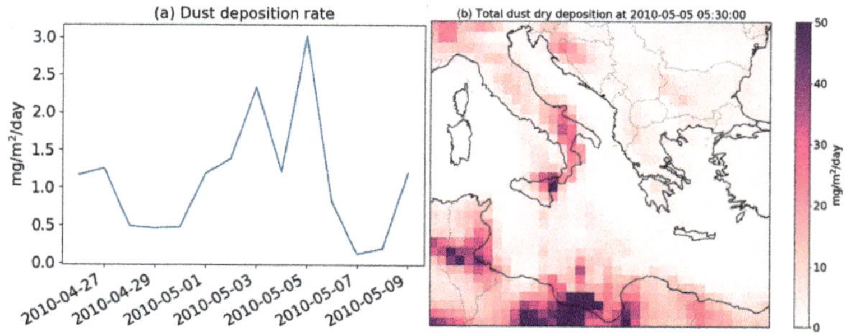

Figure 4. MARRES-2 reanalysis using GEOS-5 of dust deposition at Zagreb area from 24 April to 5 May in 2010 with visible maximum of dust on 5 May (**a**), and dust deposition over Mediterranean area, including Croatia, i.e., Zagreb area, at 5 May 2010 at 05:30. GEOS-5 system [63] has a resolution of 50 km, and forecast the dust, sea salt, POC, black carbon, and sulfates [64].

The more alkaline pH value is probably due to the neutralization of acidity in precipitation by alkaline compounds from Saharan material (i.e., increased values of Ca^{2+} and NH^{4+} according to data of DHZM). In comparison, the pH values in other rainwater samples ranged from 3.8 to 7.25 (on average, pH = 5.4), with the contribution of acid rain (pH < 5.6) of 65.5%. The higher levels of POC observed in spring rain samples might be related to the presence of biogenic particles such as pollen. Therefore, an important characteristic of the chemical composition of rainwater in the Zagreb area is episodic variation in the POC concentration and pH values as well.

Total organic carbon (TOC) was calculated as the sum of POC and DOC fractions and varied from 0.10 to 6.44 mgC L^{-1} (average, 2.05 ± 1.49). The POC contribution to TOC was dominant (about 58%) only in the case of African dust events. For all other samples, the proportion of DOC to TOC was more significant and ranged from 61 to 97% (average, 83%). According to literature, DOC makes up between 65 and 99% of TOC depending on the location and emissions [8] (and references therein). In this study, the presence of significant quantities of POC (up to 39%) may reflect a relatively high contribution of dry atmospheric OC deposition [8].

3.2. *Surface Active Reactivity Characterization of DOC in Rainwater*

The surface reactivity of the DOC was evaluated by the determination of SAS by the electrochemical method. In the present study, the relative concentrations of SAS values in rainwater samples (N = 21) varied between 0.030 and 0.358, with an average of 0.156 ± 0.10 mg L^{-1} eq. Triton-X-100. The obtained results were compared to those from previous studies in Zagreb [21,22] as well as to levels of SAS in the surface freshwater systems in Croatia (Table 2).

Table 2. Concentration of SAS (mg L^{-1} eq. Triton-X-100) in surface and rainwaters in Croatia.

Aquatic system (Croatia)	SAS	References
Rainwater (Zagreb)	0.090–0.37	[21]
Rainwater (Zagreb)	0.055–0.450 (0.140 *)	[22]
Lake Brljan (Krka River)	0.018–0.150	[57]
River Sava	0.153–0.255	[58]
River Lonja	0.209–0.258 [a]	
Lake Maksimir (Zagreb)	0.263 [b]	
Rainwater (Zagreb)	0.030–0.358 (0.156 *) [c]	

* Average value. [a] Data 2017–2019; [b] Data 2017; [c] Data 2009–2011.

As evident, the more adsorbable SAS were measured in the samples of rainwater than in the surface waters. This may be due to the presence of a higher amount of SAS in the samples but also due to the difference in their nature i.e., surface reactivity [55,65]. Namely, in freshwater systems, surface-active material corresponds mainly to humic-type substances [60], and the surface activity of atmospheric humic substances (HULIS) was found to be higher than that determined for aqueous humic materials [29]. Our SAS results probably reflect these facts. On the other hand, in the literature, it has been shown the importance of adsorption of atmospheric hydrophobic gaseous (volatile and semivolatile organic compounds, VOCs and SOCs) to the surface of atmospheric droplets (cloud, rain, and fog) i.e., on the interface of gaseous phase—atmospheric liquid phase [62,66,67]. For example Wang et al. [67] observed that concentrations of hydrophobic VOCs in cloud water can be supersaturated by a factor of $10–10^3$ in comparison with Henry's law of equilibrium. Therefore, the relatively high surface-active properties of present organic matter in atmospheric precipitation are probably a consequence of the higher concentration of hydrophobic organic contaminants in rain droplets as has been shown in our earlier studies [23,65].

In order to make a rough characterization of the SAS in the studied samples based on the different hydrophobicity, i.e., different adsorption properties at the mercury surface, the simple electrochemical method developed within the group was applied. In such an approach, the type of OM was estimated by comparison of the rain sample SAS normalized to its DOC content (called normalized surfactant activity, NSA) with the same properties (NSA) of the model substances [65]. For example, hydrophobic substances such as anionic surfactant, sodium dodecyl benzene sulfonate (NaDBS), and fulvic acid are characterized by a higher NSA, i.e., SAS/DOC ratios (0.21 and 0.17) than more hydrophilic OM with the lower SAS/DOC ratio (NSA = 0.05 for 3-hydroxybutanoic acid, 3-HBA, and 0.015 for capric acid) [65]. The application of the method is illustrated in Figure 5.

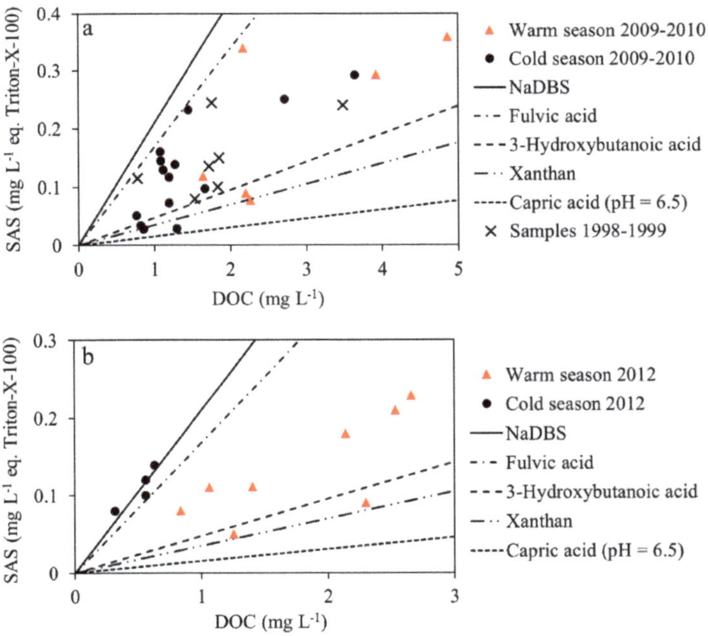

Figure 5. Correlation concentrations of SAS vs. DOC values obtained in the cold/warm season in daily rain samples in Zagreb: (**a**) bulk samples from 2009/2010 (current study) (**b**) wet—only samples from 2012 [68] The lines correspond to the model substances [65].

From Figure 5a, it can be seen that the seasonal differences (warm season, April–June, and cold season, October–March) in the type of SAS are not clearly expressed. Although data from the warm season are slightly shifted toward more hydrophilic substances, the results from the cold season fall between hydrophobic/hydrophilic behavior. The SAS/DOC ratios for precipitation in the previous investigation (1998–1999) are also distributed between fulvic acid and 3-HBA and reflect the very similar hydrophilic/hydrophobic nature of the studied OM. In daily wet-only rain samples (collected in the area of Zagreb in 2012), a defined seasonal pattern of SAS was observed, Figure 5b [68]. During autumn/winter, hydrophobic SAS was the principle fraction, while the hydrophilic fraction becomes more evident during spring/summer. Data from the cold season show a very similar adsorption (reactivity) behavior to NaDBS, while the data from the warm season are grouped close to the line for 3-HBA, pointing to the similar nature of the existing surface-active material. Taking into account that the largest fraction of hydrophobic organic material in the rain comes from incompletely combusted fossil fuels [69], the increase of fuel consumption in the heating season is probably contributing to the obtained differences in the SAS properties.

Additionally, results from an investigation on SAS in the water-soluble organic carbon (WSOC) fraction of PM_{10}, collected in the northern part of Zagreb (2011–2012), have shown a defined seasonal pattern [70]. The concentrations of WSOC fraction were lower in spring/summer (average: 2.97 ± 0.81 and 2.64 ± 0,87 μgC m^{-3}) than in autumn/winter (average: 6.43 ± 3.52 and 7.08 ± 3.16 μgC m^{-3}). During the cold season, hydrophobic WSOC was the principle fraction (autumn/winter), while the hydrophilic fraction becomes more critical during the warm season (spring/summer). A similar result was observed in atmospheric aerosols in Beijing where the mean concentrations of WSOC were 10.2 μg m^{-3} with increased values in winter and lower values in summer [71].

The complexation of copper ions (CuCC) in rainwater obtained in 11 unfiltered samples varied between 48–370 nM (Figure S1 in Supplementary Materials) with apparent stability constants (log Kapp) in the range between 6.0 and 9.5. The complexation stability constant (log Kapp) can be used to compare the relative stability of the Cu complexes [34]. In the previous study conducted in Zagreb on other location, on the hill (northern part of Zagreb) surrounded by a forest, approximately 2 km (airline) from the city center, with low traffic, the CuCC was higher, up to 586 nM with stability constants, of up to 10 [22]. It is possible that biogenic emissions contributed to these results.

The study of RSS was performed by voltammetric measurements at the Hg electrode. The presence of two different types of RSS that have different interactions with the Hg was confirmed: tiol-type, which forms a complex with Hg, i.e., RS-Hg and sulfide/S^0-like compounds which deposit the HgS [72,73]. The compounds that contribute to the RS-Hg interaction were removed by acidification and purging of the samples, indicative of their volatile nature. The acidification and purging step is usually used as an essential methodological step in the characterization of volatile and non-volatile RSS [41,72,73]. S^0-like compounds that correspond to nonvolatile RSS contributed to the HgS behavior [41,72,73]. Concentration of volatile and nonvolatile RSS in all studied precipitation samples ranged between 2–8 nM (Table S4 in Supplementary Materials) equivalent to sulfide calibration [41]. The exact nature of the detected RSS is still unknown; however, the fact that a large percentage of the rainwater DOC has been attributed to atmospheric HULIS [27,74], we assume that detected RSS can be associated with the same organic material. The oligomers, organosulfates, and nitrooxy organosulfates obtained in the atmospheric samples could play a significant part in the HULIS fraction of atmospheric organic matter [27] and could contain electrochemically active sulfur as already shown for the same type of organic materials isolated from the lagoon and marine sediments [72,75]. Such complex organic material could contribute to higher solubility and stabilization of volatile RSS in the atmospheric aqueous-phase, as is similarly obtained for natural organic macromolecules and S^0 in the seawater and water-soluble fraction of marine aerosol particles [72,73]. Sulfur-containing compounds in the class of carbohydrates and proteins have been recently identified in HULIS extracts of atmospheric aerosol particles collected in the coastal area of South Korea [76].

The same compounds are shown to be electroactive on the Hg [77,78]. This type of OM that includes RSS might be important for speciation as well as for the solubility of trace metals in the atmospheric aqueous-phase.

3.3. *Metal Content in Rainwater*

The trace metals are another important component of rainwater. The levels of twenty metals (Al, Fe, Zn, Cr, Ba, Mn, Cu, Sr, Ti, Pb, V, Ni, Rb, Sb, As, Sn, Se, Co, Cd, Mo) were investigated in rainwater samples (N = 19) in Zagreb, from April 2009 to May 2010. Table S5 in Supplementary Materials represents the descriptive statistic parameters of the total concentrations of TEs. The high standard deviation (SD) reflects the variability of trace metal concentrations. The soil elements (Al and Fe) display the maximum variability of concentrations (high SD). The lowest SD (<1) is calculated for Sn, Co, Cd, Se, and Mo. The distinct variability of different metals between different rainfalls reflects their different origins. The rainfall amount might also influence the temporal variability of metal concentrations [79]. In this study, except for V, all the other metals concentrations are significantly ($p < 0.05$) negatively correlated with the amount of precipitation, Table S6 in Supplementary Materials. Considering that the abrasion of asphalt is one of the possible sources of V [38]. Zn and Cu have the strongest correlation ($r < -0.80$; $p < 0.0001$) with the amount of rainwater. It is possible that the concentrations of these elements were associated with the seasonal variability in precipitation. Such an observation suggests that washout is the main mechanism that incorporates those traces of metals into rainwater [45].

In order to compare our results with those previously published, the average daily fluxes of metals (μg m^{-2} day^{-1}) were calculated from data regarding metal concentration and daily amount of precipitation (DHZM Data), Figure 6. The highest average fluxes (569.4 and 405.8 μg m^{-2} day^{-1}) were observed for crustal elements (Al and Fe). Also, the dominant metals in the atmospheric deposition (approximately between 20–40 μg m^{-2} day^{-1}) were Cr, Zn, Ba, Mn, while Se, Cd, Co, Mo belong to the group with the lowest deposition level (<0.4 μg m^{-2} day^{-1}). The Zn enrichment of rainwater has been observed worldwide [45] (and references therein). However, we have no explanation for high concentrations of Cr in our samples (up to 23.28 μg L^{-1} or up to 30.75 μg m^{-2} day^{-1} (Table S5 in SI, and Figure 5). There are many possible sources of Cr, such as fossil fuel combustion, resuspended surface dust, and brake wear [37,38]. The concentration of Cr in rainwater is commonly in the range from 0.2–1 μg L^{-1} [80]. It was found that the deposition of Cr occurs primarily by dry deposition, while wet deposition can only scavenge a relatively small amount of Cr from the atmosphere [81]. Considering that there is no local industrial emission that actively contributes to the abundance of the atmospheric metals in Zagreb, it can be assumed that our sampling site was contaminated by Cr.

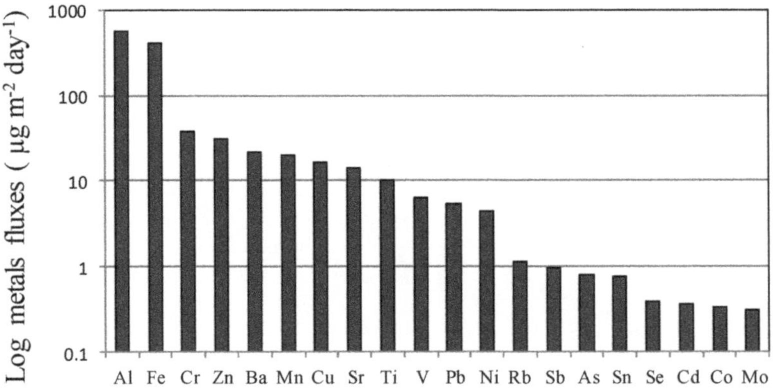

Figure 6. Average daily fluxes of trace elements (μg m^{-2} day^{-1}) in Zagreb (April 2009 to May 2010).

We compared our data with literature data using similar sampling procedures. The daily fluxes of Zn, Cu, Mn, Pb, V, and Ni were lower in Zagreb by a factor of about 4, 6, 3, 11, 9, and 7 than those at an industrial site in Belgrade [82]. The comparison was also made with data from the city of Venice [13]. Our results have shown that averaged fluxes of Zn, Ti, Pb, Ni, As, and Cd was significantly reduced (for 3, 11, 3, 2, 2, and 2 times), while data for Cu and V were comparable to those in Venice. In a similar study, conducted in the Northern Spanish coastal urban area, with low to middle pollution level [47], the daily atmospheric deposition of anthropogenic elements V, Ni, As, and Cd, were somewhat lower (by factor 2.0; 2.2; 2.6; 3.5) than those detected in Zagreb. This could be the consequence of a minor influence of industrial emissions over Zagreb, and the main sources of TMs could be ascribed to the domestic heating, traffic, and thermal power plants.

3.3.1. Sources of TEs

Correlations between elemental concentrations in rainwater can help to identify their sources in the atmosphere. The correlation matrixes, including Spearman coefficient values ($p \leq 0.05$), were given in Table S7 in Supplementary Materials. With the exception of Cr and V, all the other metals show a significant correlation between themselves. Vanadium significantly correlated (r = 0.67) only with Ni (data not shown), which is indicative of their common source—fossil-fuel combustion. It is known that V and Ni are considered as a tracer of diesel combustion [83]. Cr is not significantly correlated with Pb, Rb, As, Sn, Cd, but it shows a significant and weak correlation with the other investigated metals ($0.50 < r < 0.68$) (data not shown).

The major crustal element, Al, is highly correlated ($r > 0.90$) with Fe, Ba, Sr, Ti, Zn, and Co. Of these elements, Fe, Ba, Sr, Ti belong to the same group associated with possible natural geological sources. However, the high correlation of Zn and Co with typical geological marker elements (Al, Fe, Sr, Ti, Mn, Sr) suggests a contribution from road dust, which is enriched in these elements. Previous studies showed that the primary sources of Zn in the urban atmosphere are exhaust emissions from both gasoline and diesel-fueled road vehicles, as well as tire wear [37,80,84] (and references therein). Our result confirms this finding; Zn had a good correlation with a group of metals that generally originate from road traffic emissions (Cu, Pb, Sb, As, Co, Mo) [37,38]. Co, Mo, and Mn are correlated between themselves (Table S7 in Supplementary Materials). These elements are constituents of steel alloys and are therefore correlated [38]. There is also a very high correlation coefficient (r = 0.9) between Cd and Sb, which is mainly derived from brake linings [84]. For Cu, the highest correlation was found with Mo and As, elements that attribute to fossil fuel combustion [37]. It was previously suggested that the vehicle-derived metal Ba is mainly derived from brake wear rather than from combustion [84]. In rainwater samples, Ba correlated well with Sb and with fossil fuel combustion-generated elements such as Zn, Co, and Mo [37]. On the other hand, Ba also correlated well with crustal elements (Al, Fe, Sr, Ti, Mn). However, a significant correlation between the two analyzed concentrations does not necessarily indicate that they originate from the same type of source, but that they under certain physical conditions show similar behavior.

Enrichment factor (EF) as a good indicator for classifying the source of metals (natural or anthropogenic), it was calculated according to Duce et al. [85]. Al is selected as a reference element because it is assumed that it originates exclusively from crustal materials. Elemental crust values were found in Rudnick and Gao [86]. The average EF for metals in rainwater are plotted in Figure 7.

The elements with EF value close to 1, have a crustal source [86]. In our rainwater samples, only Ti has crustal input. Distinctly high EF values (>100) were obtained for Cd, Sb, and Cu point to their anthropogenic sources [79]. The high EF for these elements reflect the local emissions from traffic activities (brake wear). The values of EF in the range between 1 and 10 might indicate the influence of local soil composition [86]. As can be seen from Figure 6, Fe, Rb, Co, Se, Ba, V, and Sr belong to this group. The remaining metals (Ni, As, Mo Sn, Mn, Pb, Cr, Zn), having EF values in the range of 10 to 100 belong to categories classified as moderately enriched, suggesting other sources in addition to crustal material [86].

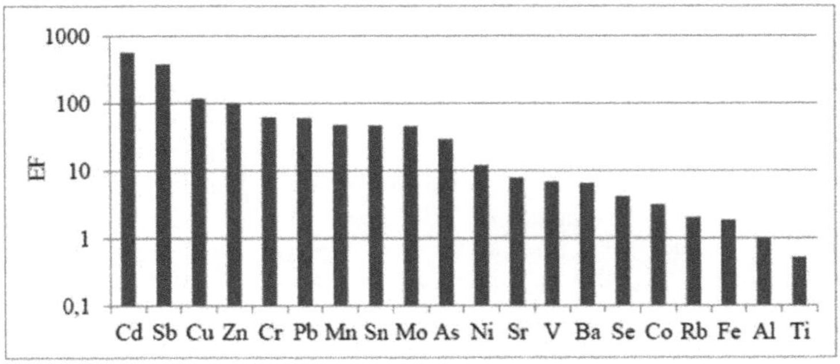

Figure 7. Average enrichment factors (EF) for trace elements in rainwater.

3.3.2. Solubility of TEs in Rainwater

The solubility of TEs determines the distribution of pollutants and bioavailability of deposition [39]. In order to discuss the dissolution tendency of TEs in deposition, we measured the metal concentrations in the 14 filtered samples. In this case, the soluble fraction was defined as that passing through a 0.45 μm membrane filter. For each considered trace metal, the percentage of the solubility was calculated as the ratio [filtered fraction/unfiltered (%)]. The results (average solubility) of all samples are presented in Figure 8.

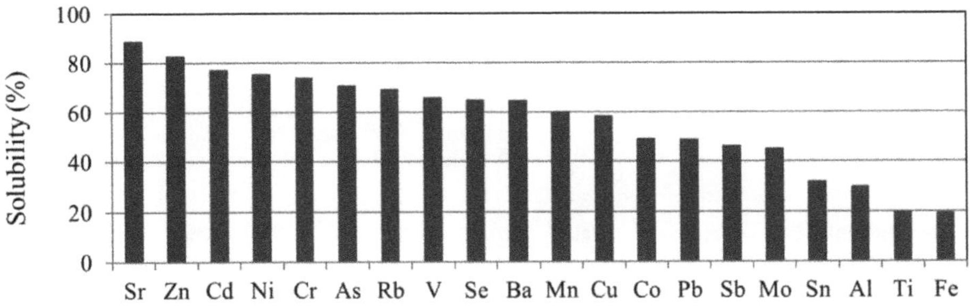

Figure 8. Average solubility of the trace metals in rainwater.

The average solubility of metals was between 19 and 89%. The solubility of Sr, Zn, Cd, Ni, Cr, As, and Rb exhibits high values, generally greater than 70%. Sr and Zn (>80%) showed the highest solubility. The solubilities of Al, Sn, Mo, Sb, Pb, Co, Cu, Mn, Ba, Se, V, and Rb ranged from 30 to 70%. The crustal elements Al, Ti, and Fe, have the lowest solubility (below 20%), which is in line with the existing literature [44,46,87].

In the literature, it has been shown that pH is an important factor for the solubility of some metals [31,48] (and references therein). However, in this study, no correlation was found between pH values and the solubility of metals ($p > 0.05$). Probably, in the bulk samples, the relationship between pH values and solubility of metals is masked. Similar results have been obtained in the study of atmospheric bulk deposition in Paris [88].

4. Conclusions

In this study, in daily bulk precipitation samples collected in the urban area of Croatia (Zagreb city center), the characteristics of OM and TEs were investigated. This investigation included measurements of the magnitude and variability of the DOC, POC, and TEs, as well as of the selected reactive classes of organic and sulfur compounds (SAS, CuCC, RSS), having a critical function in both the atmospheric processes and natural surface waters.

The concentrations (0.69–4.86 mg L^{-1}) and the reactivity of DOC (estimated based on the SAS content) demonstrated that the level and composition of DOC in this area had not changed significantly during the observed decade (1998–1999 versus 2009–2011). POC concentrations in rainwater sampled during 2009–2010 varied. The concentration of POC was the most significantly affected by the Saharan dust event.

The study of TEs in rainwater has shown that deposition fluxes of the metals were lower than those published in the literature for the urban areas of Europe. Dominant metals in the atmospheric deposition were Al and Fe (crustal elements). Enrichment factors show that rainwater samples were moderately enriched with Ni, As, Mo Sn, Mn, Pb, Cr, Zn. The very high EF calculated for Cd, Sb, and Cu reflect their anthropogenic origin. The high average solubility (>70%) was found for Sr, Zn, Cd, Ni, Cr, A, and Rb, while the lowest solubility (below 20%) for crustal elements (Al, Fe, and Ti).

Results from this study clearly show how monitoring of precipitation could be an important factor for evaluating the air quality in an urban area and for estimation of atmospheric impact on the other ecosystems. At the same time, the physico-chemical characterization of the rainwater can be a base for a better insight into complex atmospheric aqueous chemistry as well. In addition, this unique study for Zagreb, and Croatia, could serve as a baseline for future investigations in the area.

Supplementary Materials: The following are available online at http://www.mdpi.com/2073-4433/11/2/144/s1, Figure S1: Copper complexing capacity (CuCC) in precipitation., Table S1: Performance of HR ICP-MS analyses: quality control (QC) as represented by analyzing certified standard reference materials "River Water Reference Material for Trace Metals" (SLRS-4, National Research Council Canada) (n = 11), the common level of procedural blanks and method Level of Quantification (LOQ). All values are provided in μg/L, Table S2: Concentration of DOC and pH values in rainwater (Zagreb). Table S3: Correlation between DOC and metal concentrations in rainwater (Zagreb). Tabel S4: RSS concentration and type of interaction with Hg in rainwater samples. Tabel S5: Arithmetic means (μg/L), standard deviations (SD), median, minimum, maximum of trace elements in bulk samples (N = 19) of rainwater. Tabel S6: The relationship between metal concentrations and rainfall amounts. (Spearman correlation coefficients, r and *p* values). Tabel S7: Spearman correlation matrix for the metals in rainwater samples.

Author Contributions: P.O.-L.: initial study idea, original draft writing. K.V.: data processing for TEs and reviewing the manuscript. I.C.: RSS measurements, initial draft conceptualization and discussion, funding. D.O.: trace metals measurements. I.Š.: data collection, reviewing the manuscript. M.D.S.: MARRES-2 reanalysis using GEOS-5 of dust deposition at Zagreb area. All authors have read and agreed to the published version of the manuscript.

Funding: This research received no external funding.

Acknowledgments: Croatian Science Foundation project IP-2018-01-1717, MARRES is highly acknowledged for unifying of all presented results in this paper. The authors thank Z. Zovko and J. Dautović for the DOC and POC measurements and M. Plavšić for the CuCC measurements, while A. Penzić and S. Frka are highly acknowledged for collecting and SAS measurments in daily wet-only rain samples in 2012. DHZM (I Igrec) is acknowledged for providing rainfall data.

Conflicts of Interest: The authors declare no conflict of interest

References

1. Brimblecombe, P. *Air Composition and Chemistry*; Cambridge University Press: Cambridge, UK, 1996.
2. Willey, J.D.; Glinski, D.A.; Southwell, M.; Long, M.S.; Brooks Avery, G.; Kieber, R.J. Decadal variations of rainwater formic and acetic acid concentrations in Wilmington, NC, USA. *Atmos. Environ.* **2011**, *45*, 1010–1014. [CrossRef]
3. Jeong, G.Y.; Kim, J.Y.; Seo, J.; Kim, G.M.; Jin, H.C.; Chun, Y. Long-range transport of giant particles in Asian dust identified by physical, mineralogical, and meteorological analysis. *Atmos. Chem. Phys.* **2014**, *14*, 505–521. [CrossRef]
4. Richon, C.; Dutay, J.C.; Dulac, F.; Wang, R.; Balkanski, Y. Modeling the biogeochemical impact of atmospheric phosphate deposition from desert dust and combustion sources to the Mediterranean Sea. *Biogeosciences* **2018**, *15*, 2499–2524. [CrossRef]
5. Willey, J.D.; Kieber, R.J.; Eyman, M.S.; Avery, G.B., Jr. Rainwater dissolved organic carbon: Concentrations and global flux. *Glob. Biogeochem. Cycles* **2000**, *14*, 139–148. [CrossRef]
6. Canic, K.S.; Vidic, S.; Klaic, Z.B. Precipitation chemistry in Croatia during the period 1981–2006. *J. Environ. Monit.* **2009**, *11*, 839–851. [CrossRef] [PubMed]

7. Špoler Čanić, K.K., I; Bencetić Klaić, Z. Impact of Saharan Dust on Precipitation Chemistry in Croatia. Available online: https://citeseerx.ist.psu.edu/viewdoc/summary?doi=10.1.1.680.5060 (accessed on 13 October 2019).
8. Iavorivska, L.; Boyer, E.W.; DeWalle, D.R. Atmospheric deposition of organic carbon via precipitation. *Atmos. Environ.* **2016**, *146*, 153–163. [CrossRef]
9. Iavorivska, L.; Boyer, E.W.; Grimm, J.W. Wet atmospheric deposition of organic carbon: An underreported source of carbon to watersheds in the northeastern United States. *J. Geophys. Res. Atmospheres* **2017**, *122*, 3104–3115. [CrossRef]
10. Bao, H.Y.; Yi, Y.Y.; Wang, C.; Spencer, R.G.M.; Deng, X.; Guo, W.D. Dissolved organic matter in coastal rainwater: Concentration, bioavailability and depositional flux to seawater in southeastern China. *Mar. Chem.* **2018**, *205*, 48–55. [CrossRef]
11. Lawlor, A.; Tipping, E. Metals in bulk deposition and surface waters at two upland locations in northern England. *Environ. Pollut.* **2003**, *121*, 153–167. [CrossRef]
12. Tuncel, G.W.a.d.d.f.o.m.i.i.t.E.M. Black Sea and Central Anatolia. *Map Tech. Rep. Series* **2001**, *133*, 1–34.
13. Rossini, P.; Guerzoni, S.; Molinaroli, E.; Rampazzo, G.; De Lazzari, A.; Zancanaro, A. Atmospheric bulk deposition to the lagoon of Venice - Part I. Fluxes of metals, nutrients and organic contaminants. *Environ. Int.* **2005**, *31*, 959–974. [CrossRef] [PubMed]
14. Chow, M.F.; Lai, C.C.; Kuo, H.Y.; Lin, C.H.; Chen, T.Y.; Shiah, F.K. Long Term Trends and Dynamics of Dissolved Organic Carbon (DOC) in a Subtropical Reservoir Basin. *Water* **2017**, *9*, 14. [CrossRef]
15. Kieber, R.J.; Peake, B.; Willey, J.D.; Avery, G.B. Dissolved organic carbon and organic acids in coastal New Zealand rainwater. *Atmos. Environ.* **2002**, *36*, 3557–3563. [CrossRef]
16. Pan, Y.; Wang, Y.; Xin, J.; Tang, G.; Song, T.; Wang, Y.; Li, X.; Wu, F. Study on dissolved organic carbon in precipitation in Northern China. *Atmos. Environ.* **2010**, *44*, 2350–2357. [CrossRef]
17. Siudek, P.; Frankowski, M.; Siepak, J. Seasonal variations of dissolved organic carbon in precipitation over urban and forest sites in central Poland. *Environ. Sci. Pollut. Res.* **2015**, *22*, 11087–11096. [CrossRef] [PubMed]
18. Yan, G.; Kim, G. Dissolved organic carbon in the precipitation of Seoul, Korea: Implications for global wet depositional flux of fossil-fuel derived organic carbon. *Atmos. Environ.* **2012**, *59*, 117–124. [CrossRef]
19. Montero-Martínez, G.; Rivera-Arellano, J.; Roy, P.D.; Rosado-Abón, A.; Hernández-Nagay, D.P.; Mendoza-Trejo, A.; Andraca-Ayala, G.L. Content and composition of dissolved organic carbon in precipitation at the southern part of Mexico City. *Atmosfera* **2018**, *31*, 331–346. [CrossRef]
20. Niu, H.; Kang, S.; Shi, X.; Zhang, G.; Wang, S.; Pu, T. Dissolved organic carbon in summer precipitation and its wet deposition flux in the Mt. Yulong region, southeastern Tibetan Plateau. *J. Atmos. Chem.* **2019**, *76*, 1–20. [CrossRef]
21. Leko, P.O.; Kozarac, Z.; Ćosović, B. Surface active substances (SAS) and dissolved organic matter (DOC) in atmospheric precipitation of urban area of Croatia (Zagreb). *Water Air Soil Pollut.* **2004**, *158*, 295–310. [CrossRef]
22. Orlović-Leko, P.; Plavšić, M.; Bura-Nakić, E.; Kozarac, Z.; Ćosović, B. Organic matter in the bulk precipitations in Zagreb and Šibenik, Croatia. *Atmos. Environ.* **2009**, *43*, 805–811. [CrossRef]
23. Orlović-Leko, P.; Kozarac, Z.; Ćosović, B.; Strmečki, S.; Plavšić, M. Characterization of atmospheric surfactants in the bulk precipitation by electrochemical tools. *J. Atmos. Chem.* **2010**, *66*, 11–26. [CrossRef]
24. Renard, P.; Canet, I.; Sancelme, M.; Wirgot, N.; Deguillaume, L.; Delort, A.M. Screening of cloud microorganisms isolated at the Puy de Dome (France) station for the production of biosurfactants. *Atmos. Chem. Phys.* **2016**, *16*, 12347–12358. [CrossRef]
25. Ćosović, B. Adsorption kinetics of the complex mixture of organic solutes at model and natural phase boundaries. In *Aquatic Chemical Kinetics, Environmental Science and Technology Series*; Weily: New York, NY, USA, 1990; pp. 291–311.
26. Seidl, W. Model for a surface film of fatty acids on rain water and aerosol particles. *Atmos. Environ.* **2000**, *34*, 4917–4932. [CrossRef]
27. Graber, E.; Rudich, Y. Atmospheric HULIS: How humic-like are they? A comprehensive and critical review. *Atmos. Chem. Phys.* **2006**, *6*, 729–753. [CrossRef]
28. Kristensen, T.B.; Du, L.; Nguyen, Q.T.; Nojgaard, J.K.; Koch, C.B.; Nielsen, O.F.; Hallar, A.G.; Lowenthal, D.H.; Nekat, B.; van Pinxteren, D.; et al. Chemical properties of HULIS from three different environments. *J. Atmos. Chem.* **2015**, *72*, 65–80. [CrossRef]

29. Santos, P.S.; Santos, E.B.; Duarte, A.C. First spectroscopic study on the structural features of dissolved organic matter isolated from rainwater in different seasons. *Sci. Total Environ.* **2012**, *426*, 172–179. [CrossRef] [PubMed]
30. Willey, J.D.; Kieber, R.J.; Seaton, P.J.; Miller, C. Rainwater as a source of Fe(II)-stabilizing ligands to seawater. *Limnol. Oceanogr.* **2008**, *53*, 1678–1684. [CrossRef]
31. Okochi, H.; Brimblecombe, P. Potential trace metal–organic complexation in the atmosphere. *The Sci. World J.* **2002**, *2*, 767–786. [CrossRef]
32. Witt, M.; Jickells, T. Copper complexation in marine and terrestrial rain water. *Atmos. Environ.* **2005**, *39*, 7657–7666. [CrossRef]
33. Witt, M.; Skrabal, S.; Kieber, R.; Willey, J. Copper complexation in coastal rainwater, southeastern USA. *Atmos. Environ.* **2007**, *41*, 3619–3630. [CrossRef]
34. Plavšić, M.; Orlović-Leko, P.; Kozarac, Z.; Bura-Nakić, E.; Strmečki, S.; Ćosović, B. Complexation of copper ions in atmospheric precipitation in Croatia. *Atmos. Res.* **2008**, *87*, 80–87. [CrossRef]
35. Louis, Y.; Garnier, C.; Lenoble, V.; Mounier, S.; Cukrov, N.; Omanovic, D.; Pizeta, I. Kinetic and equilibrium studies of copper-dissolved organic matter complexation in water column of the stratified Krka River estuary (Croatia). *Mar. Chem.* **2009**, *114*, 110–119. [CrossRef]
36. Baker, A.R.; Croot, P.L. Atmospheric and marine controls on aerosol iron solubility in seawater. *Mar. Chem.* **2010**, *120*, 4–13. [CrossRef]
37. Allen, A.; Nemitz, E.; Shi, J.; Harrison, R.; Greenwood, J. Size distributions of trace metals in atmospheric aerosols in the United Kingdom. *Atmos. Environ.* **2001**, *35*, 4581–4591. [CrossRef]
38. Johansson, C.; Norman, M.; Burman, L. Road traffic emission factors for heavy metals. *Atmos. Environ.* **2009**, *43*, 4681–4688. [CrossRef]
39. Faloona, I. Sulfur processing in the marine atmospheric boundary layer: A review and critical assessment of modeling uncertainties. *Atmos. Environ.* **2009**, *43*, 2841–2854. [CrossRef]
40. Watts, S.F. The mass budgets of carbonyl sulfide, dimethyl sulfide, carbon disulfide and hydrogen sulfide. *Atmos. Environ.* **2000**, *34*, 761–779. [CrossRef]
41. Orlović-Leko, P.; Omanović, D.; Ciglenečki, I.; Vidović, K.; Brenko, T. Application of electrochemical methods in the physicochemical characterization of atmospheric precipitation. *Bulg. Chem. Commun.* **2017**, *49*, 211–217.
42. Rozan, T.F.; Lassman, M.E.; Ridge, D.P.; Luther, G.W., III. Evidence for iron, copper and zinc complexation as multinuclear sulphide clusters in oxic rivers. *Nature* **2000**, *406*, 879. [CrossRef]
43. Wong, C.S.C.; Li, X.; Thornton, I. Urban environmental geochemistry of trace metals. *Environ. Pollut.* **2006**, *142*, 1–16. [CrossRef]
44. Al-Momani, I.F. Trace elements in atmospheric precipitation at Northern Jordan measured by ICP-MS: Acidity and possible sources. *Atmos. Environ.* **2003**, *37*, 4507–4515. [CrossRef]
45. Hou, H.; Takamatsu, T.; Koshikawa, M.; Hosomi, M. Trace metals in bulk precipitation and throughfall in a suburban area of Japan. *Atmos. Environ.* **2005**, *39*, 3583–3595. [CrossRef]
46. Okubo, A.; Takeda, S.; Obata, H. Atmospheric deposition of trace metals to the western North Pacific Ocean observed at coastal station in Japan. *Atmos. Res.* **2013**, *129*, 20–32. [CrossRef]
47. Fernández-Olmo, I.; Puente, M.; Montecalvo, L.; Irabien, A. Source contribution to the bulk atmospheric deposition of minor and trace elements in a Northern Spanish coastal urban area. *Atmos. Res.* **2014**, *145*, 80–91. [CrossRef]
48. Başak, B.; Alagha, O. Trace metals solubility in rainwater: Evaluation of rainwater quality at a watershed area, Istanbul. *Environ. Monit. Assess.* **2010**, *167*, 493–503. [CrossRef] [PubMed]
49. Heimburger, A.; Losno, R.; Triquet, S. Solubility of iron and other trace elements in rainwater collected on the Kerguelen Islands (South Indian Ocean). *Biogeosciences* **2013**, *10*, 6617–6628. [CrossRef]
50. Statistics Cost-Based Optimizer. Statistical Yearbook of the Republic of Croatia. Available online: http://www.dzs.hr/Hrv_Eng/ljetopis/2010/SLJH2010.pdf (accessed on 13 October 2019).
51. Statistics Cost-Based Optimizer. Statistical Yearbook of the Republic of Croatia. Available online: http://www.dzs.hr/Hrv_Eng/ljetopis/2011/SLJH2011.pdf (accessed on 12 September 2019).
52. Statistical Yearbook of the Republic of Croatia. Available online: https://www.dzs.hr/Hrv_Eng/ljetopis/2012/sljh2012.pdf (accessed on 12 September 2019).

53. Kanakidou, M.; Seinfeld, J.H.; Pandis, S.N.; Barnes, I.; Dentener, F.J.; Facchini, M.C.; Van Dingenen, R.; Ervens, B.; Nenes, A.; Nielsen, C.J. Organic aerosol and global climate modelling: A review. *Atmos. Chem. Phys.* **2005**, *5*, 1053–1123. [CrossRef]
54. Cosovic, B.; Vojvodic, V. Voltammetric analysis of surface active substances in natural seawater. *Electroanalysis* **1998**, *10*, 429–434. [CrossRef]
55. Orlović-Leko, P.; Vidović, K.; Plavšić, M.; Ciglenečki, I.; Šimunić, I.; Minkina, T. Voltammetry as a tool for rough and rapid characterization of dissolved organic matter in the drainage water of hydroameliorated agricultural areas in Croatia. *J. Solid State Electrochem.* **2016**, *20*, 3097–3105. [CrossRef]
56. Balla, D.; Papageorgiou, A.; Voutsa, D. Carbonyl compounds and dissolved organic carbon in rainwater of an urban atmosphere. *Environ. Sci. Pollut. Res.* **2014**, *21*, 12062–12073. [CrossRef]
57. Strmecki, S.; Ciglenecki, I.; Udovic, M.G.; Margus, M.; Bura-Nakic, E.; Dautovic, J.; Plavsic, M. Voltammetric Study of Organic Matter Components in the Upper Reach of the Krka River, Croatia. *Croat. Chem. Acta* **2018**, *91*, 447–454. [CrossRef]
58. Šimunić, I.; Orlović-Leko, P.; Marić, A.S.; Ciglenečki, I.; Tatiana, M.; Filipović, V. Quality of Surface Water in the Agricultural District Lonja Field (Croatia). *Agric. For.* **2017**, *63*, 5–14.
59. Orlović-Leko, L.Š.I.; DautovićOrlović-Leko, J.; Ciglenečki, I. Dissolved Organic Carbon (DOC) in the surface waters in the agricultural area (Lonja field, Croatia). In *7th Croatian Water Conference, Croatian Waters in Environmetal and Nature Protection*; Biondić, D., Holjević, D., Vizner, M., Eds.; Opatija, Croatia, 2019; pp. 287–291.
60. Cvrković-Karloci, Ž.; Krznarić, D.; Šeruga, M.; Ćosović, B. Simple electrochemical determination of surface-active substances in natural waters. *Int. J. Electrochem.* **2011**, *2011*. [CrossRef]
61. Cukrov, N.; Cmuk, P.; Mlakar, M.; Omanovic, D. Spatial distribution of trace metals in the Krka River, Croatia: An example of the self-purification. *Chemosphere* **2008**, *72*, 1559–1566. [CrossRef]
62. Capel, P.D.; Gunde, R.; Zuercher, F.; Giger, W. Carbon speciation and surface tension of fog. *Environ. Sci. Technol.* **1990**, *24*, 722–727. [CrossRef]
63. Buchard, V.; Randles, C.A.; da Silva, A.M.; Darmenov, A.; Colarco, P.R.; Govindaraju, R.; Ferrare, R.; Hair, J.; Beyersdorf, A.J.; Ziemba, L.D. The MERRA-2 Aerosol Reanalysis, 1980 Onward. Part II: Evaluation and Case Studies. *J. Clim.* **2017**, *30*, 6851–6872. [CrossRef]
64. Global Modeling and Assimilation Office. Available online: https://gmao.gsfc.nasa.gov/GMAO_products/reanalysis_products.php (accessed on 20 January 2020).
65. Ćosović, B.; Orlović Leko, P.; Kozarac, Z. Rainwater dissolved organic carbon: Characterization of surface active substances by electrochemical method. *Electroanal. Int. J. Devoted Fundam. Pract. Aspects Electroanal.* **2007**, *19*, 2077–2084.
66. Simcik, M.F. The importance of surface adsorption on the washout of semivolatile organic compounds by rain. *Atmos. Environ.* **2004**, *38*, 491–501. [CrossRef]
67. Wang, M.; Perroux, H.; Fleuret, J.; Bianco, A.; Bouvier, L.; Colomb, A.; Borbon, A.; Deguillaume, L. Anthropogenic and biogenic hydrophobic VOCs detected in clouds at the puy de Dôme station using Stir Bar Sorptive Extraction: Deviation from the Henry's law prediction. *Atmos. Res.* **2020**, 104844. [CrossRef]
68. Orlović-Leko, P.F.S.; Penezić, A.; Vidović, K.; Plećaš, S. Surface activity of precipitation and aerosol on the Zagreb area. In Proceedings of the 6th Croatian Water Conference, Opatija, Croatia, 20–23 May 2015.
69. Avery, G.B.; Biswas, K.F.; Mead, R.; Southwell, M.; Willey, J.D.; Kieber, R.J.; Mullaugh, K.M. Carbon isotopic characterization of hydrophobic dissolved organic carbon in rainwater. *Atmos. Environ.* **2013**, *68*, 230–234. [CrossRef]
70. Vidović, K. Seasonal Variation of Properties of Organic Matter in Aerosol in the Atmosphere over Zagreb. Master's Thesis, University of Zagreb, Zagreb, Croatian, 2014.
71. Tang, X.; Zhang, X.S.; Wang, Z.W.; Ci, Z.J. Water-soluble organic carbon (WSOC) and its temperature-resolved carbon fractions in atmospheric aerosols in Beijing. *Atmos. Res.* **2016**, *181*, 200–210. [CrossRef]
72. Bura-Nakić, E.; Helz, G.R.; Ciglenečki, I.; Ćosović, B. Seasonal variations in reduced sulfur species in a stratified seawater lake (Rogoznica Lake, Croatia); evidence for organic carriers of reactive sulfur. *Geochim. Cosmochim. Acta* **2009**, *73*, 3738. [CrossRef]
73. Cvitešić Kušan, A.F.S.; Ciglenečki, I. Electrochemical Evidence of non-Volatile Reduced Sulfur Species in Water-Soluble Fraction of Fine Marine Aerosols. *Atmosphere* **2019**, *10*, 674. [CrossRef]

74. Feng, J.; Möller, D. Characterization of water-soluble macromolecular substances in cloud water. *J. Atmos. Chem.* **2004**, *48*, 217–233. [CrossRef]
75. Gašparović, B.; Ćosović, B. Electrochemical estimation of the dominant type of surface active substances in seawater samples using o-nitrophenol as a probe. *Mar. Chem.* **1994**, *46*, 179–188. [CrossRef]
76. Jang, K.-S.; Choi, A.Y.; Choi, M.; Kang, H.; Kim, T.-W.; Park, K.-T. Size-Segregated Chemical Compositions of HULISs in Ambient Aerosols Collected during the Winter Season in Songdo, South Korea. *Atmosphere* **2019**, *10*, 226. [CrossRef]
77. Strmečki, S.; Plavšić, M.; Steigenberger, S.; Passow, U. Characterization of phytoplankton exudates and carbohydrates in relation to their complexation of copper, cadmium and iron. *Mar. Ecol. Prog. Ser.* **2010**, *408*, 33–46. [CrossRef]
78. Strmečki, S.; Paleček, E. Adsorption/desorption of biomacromolecules involved in catalytic hydrogen evolution. *Bioelectrochemistry* **2018**, *120*, 87–93. [CrossRef]
79. Al-Khashman, O.A. Study of chemical composition in wet atmospheric precipitation in Eshidiya area, Jordan. *Atmos. Environ.* **2005**, *39*, 6175–6183. [CrossRef]
80. Potential for Human Exposure. Available online: https://www.atsdr.cdc.gov/toxprofiles/tp7-c6.pdf (accessed on 7 October 2019).
81. Gunawardena, J.; Egodawatta, P.; Ayoko, G.A.; Goonetilleke, A. Atmospheric deposition as a source of heavy metals in urban stormwater. *Atmos. Environ.* **2013**, *68*, 235–242. [CrossRef]
82. Mijic, Z.; Stojic, A.; Perisic, M.; Rajsic, S.; Tasic, M.; Radenkovic, M.; Joksic, J. Seasonal variability and source apportionment of metals in the atmospheric deposition in Belgrade. *Atmos. Environ.* **2010**, *44*, 3630–3637. [CrossRef]
83. Fabretti, J.F.; Sauret, N.; Gal, J.F.; Maria, P.C.; Scharer, U. Elemental characterization and source identification of PM2.5 using Positive Matrix Factorization: The Malraux road tunnel, Nice, France. *Atmos. Res.* **2009**, *94*, 320–329. [CrossRef]
84. Sternbeck, J.; Sjodin, A.; Andreasson, K. Metal emissions from road traffic and the influence of resuspension—Results from two tunnel studies. *Atmos. Environ.* **2002**, *36*, 4735–4744. [CrossRef]
85. Duce, R.A.; Hoffman, G.L.; Zoller, W.H. Atmospheric Trace-Metals at Remote Northern and Southern-Hemisphere Sites—Pollution or Natural. *Science* **1975**, *187*, 59–61. [CrossRef]
86. Rudnick, R.L.; Gao, S. The Crust, 3.01—The Composition of the Continental Crust. In *Treatise on Geochemistry*; Elsevier-Pergamon: Oxford, UK; pp. 1–64.
87. Giusti, L.; Yang, Y.L.; Hewitt, C.N.; Hamiltontaylor, J.; Davison, W. The Solubility and Partitioning of Atmospherically Derived Trace-Metals in Artificial and Natural-Waters—A review. *Atmos. Environ. Part A Gen. Top.* **1993**, *27*, 1567–1578. [CrossRef]
88. Motelay-Massei, A.; Ollivon, D.; Tiphagne, K.; Garban, B. Atmospheric bulk deposition of trace metals to the Seine River Basin, France: Concentrations, sources and evolution from 1988 to 2001 in Paris. *Water Air Soil Pollut.* **2005**, *164*, 119–135. [CrossRef]

atmosphere

MDPI

Article

Electrochemical Evidence of non-Volatile Reduced Sulfur Species in Water-Soluble Fraction of Fine Marine Aerosols

Ana Cvitešić Kušan [1], Sanja Frka [2] and Irena Ciglenečki [1,*]

[1] Laboratory for Physical Oceanography and Chemistry of Aquatic Systems, Division for Marine and Environmental Research, Ruđer Bošković Institute, 10 000 Zagreb, Croatia; Ana.Cvitesic.Kusan@irb.hr

[2] Laboratory for Marine and Atmospheric Biogeochemistry, Division for Marine and Environmental Research, Ruđer Bošković Institute, 10 000 Zagreb, Croatia; Sanja.Frka.Milosavljevic@irb.hr

* Correspondence: irena@irb.hr

Received: 27 September 2019; Accepted: 30 October 2019; Published: 1 November 2019

Abstract: The traditional voltammetric method at the mercury electrode, and an acidification step developed for the determination of reduced sulfur species (RSS) in natural waters, was for the first time used for the quantification of RSS in the water-soluble fraction of fine marine aerosols collected at the Middle Adriatic location (Rogoznica Lake). The evidence of two types of non-volatile RSS that have different interaction with the Hg electrode was confirmed: mercapto-type which complexes Hg as RS–Hg and sulfide/S0-like compounds which deposits HgS. The analytical protocol that was used for RSS determination in aerosol samples is based on separate voltammetric studies of a methyl 3-mercaptopropionate (3-MPA) as a representative of mercapto-type compounds and sulfide as a representative of inorganic RSS. Our preliminary study indicates the presence of mainly RS–Hg compounds in spring samples, ranging from 2.60–15.40 ng m^{-3}, while both, the mercapto-type (0.48–2.23 ng m^{-3}) and sulfide and/or S0-like compounds (0.02–0.26 ng m^{-3}) were detected in early autumn samples. More expressed and defined RS–Hg peaks recorded in the spring potentially indicate their association with biological activity in the area. Those samples were also characterized by a higher water-soluble organic carbon content and a more abundant surface-active fraction, pointing to enhanced solubility and stabilization of RSS in the aqueous atmospheric phase.

Keywords: reduced sulfur species; marine aerosols; water-soluble fraction; voltammetry; mercapto-type compounds; Rogoznica Lake

---

1. Introduction

In different aquatic ecosystems, electrochemical methods have been widely used for the characterization and determination of different sulfur (S) species, comprising dissolved and/or particulate inorganic and organic S compounds, including thiols ([1–12] and references therein). Strong interaction between mercury electrode (Hg) and S species is a background for their electroanalytical determination and speciation at the Hg electrode. The methodology is based on the tendency of inorganic and organic reduced S species (RSS) to deposit a HgS layer [1–5,13–15] and/or RS–Hg complexes [9,10,15] (term "complex" refers to a different type of interaction between the analyte and the Hg electrode) during an accumulation step at the deposition potential (E_d) around −0.2 V vs. the reference electrode (Ag/AgCl). In solutions containing sulfide anions, an insoluble HgS layer is formed during the deposition step at the Hg surface by the reversible process of a two-electron-transfer oxidation of Hg at potentials more positive than −0.5 V vs. the reference electrode (Ag/AgCl) (Equation (1)) [1–15]:

$$HS^- + Hg^0 \longleftrightarrow HgS(s) + H^+ + 2e^-. \quad (1)$$

The same process occurs during the interaction of Hg with some other organic (thiourea, thiols, oxines, thioanions) and inorganic S species (polysulfides, dissolved and colloidal S, labile chalcogenides, i.e., CuS, PbS, HgS, FeS, Ag_2S). During the potential scan toward negative values, layers of HgS and adsorbable RS–Hg complexes (in the case of DMS, DMDS, cysteine, glutathione) are reduced between −0.45 and −0.70 V (vs. Ag/AgCl) with facilitated reduction of the RS–Hg complexes at a more positive potential than −0.68 V (vs. Ag/AgCl) where HgS reduction occurs [2,9,10,15,16].

Natural cycling of S compounds through the environment has taken on a new significance due to their involvement in the formation of atmospheric aerosol particles and their influence on global environmental processes and human health [17–19]. Atmospheric S contributes the most to atmospheric acid deposition being detrimental to ecosystems, harming aquatic biota, as well as to a wide range of terrestrial plant life [20]. Moreover, S-rich atmospheric particulate matter (PM) can serve as cloud condensation nuclei and participate in the cloud formation processes, which can ultimately affect the radiative balance of the atmosphere and the Earth's climate [21,22]. The main anthropogenic sources of atmospheric S are coal and oil combustion, oil refining, and smelting of copper ores ([23,24] and references therein). The widespread combustion of fossil fuels has greatly increased S emissions into the atmosphere, with the anthropogenic component now substantially greater than natural emissions on a global basis [25]. The main natural sources are oceanic phytoplankton [21], volcanoes [26,27], and geothermal fields [28]. Production of marine S compounds is mostly related to eutrophication phenomena and spreading of hypoxic–anoxic dead zones in the marine environment. A significant fraction of natural S emission occurs in the form of dimethylsulfide (DMS), produced by phytoplankton and zooplankton grazing, which is released from the ocean into the atmosphere, where it undergoes oxidation to form sulfur dioxide (SO_2), dimethylsulfoxide, dimethylsulfone, methanesulfonic acid (MSA), and sulfate (SO_4^{2-}) [29–31].

The wide range of S oxidation states (from −2 to +6) as well as organic and inorganic forms present in ambient samples makes the characterization of aerosol S challenging [32]. Sulfur in aerosols is usually measured in the form of inorganic sulfate ion, and it is often assumed to be one of the most important forms of aerosol S. However, in addition to sulfate, other S compounds, even in smaller quantities, are present in ambient aerosols. Among these, the most abundant are sulfides, organosulfur/sulfate compounds, and polycyclic aromatic sulfur heterocycles (PAHSs) [33–35]. Even though different methodological approaches have been applied to resolve S content in aerosols, including ion chromatography, X-ray fluorescence, and inductively coupled plasma mass spectrometry, these cannot determine the oxidation state or directly identify the chemical form of aerosol S. Thus, there is still a need for accurate and direct method enabling speciation of different S species in small amounts such as those found within $PM_{2.5}$ and/or PM_{10} samples (usually no more than a few milligrams).

In this work, for the first time the water-soluble (WS) fraction of the fine marine aerosols was studied by voltammetric methods that are used thus far for RSS characterization in natural waters. An analytical protocol for the electrochemical RSS characterization in the WS fraction of aerosol samples is given based on separate voltammetric studies of a methyl 3-mercaptopropionate (3-MPA), and its mixture with sulfide. Here, 3-MPA is used as a representative for mercapto-type compounds (RS–Hg type) which gives a more positive reduction peak (around −0.60 V vs. Ag/AgCl at neutral pH) than standard HgS reduction (around −0.70 V vs. Ag/AgCl) usually considered for RSS quantification in different natural samples [1–8].

2. Experiments

2.1. Materials and Methods

The chemicals used were of reagent grade. The 3-MPA Me 3-MP stock solutions were prepared by dissolving 98% methyl 3-mercaptopropionate (Merck, Darmstadt, Germany) in N_2 degassed ultra-pure water (Milli-Q, 18.2 MΩ, total organic carbon (TOC) < 3 ppb). The inorganic sulfide stock solutions were prepared by dissolving Na_2S crystals (Sigma-Aldrich, Steinheim, Germany) in N_2 degassed solution of 2×10^{-4} M NaOH (Merck, Darmstadt, Germany). Working solutions were prepared by adding, small volumes of the 3-MPA and/or Na_2S stock solutions to the previously N_2 degassed solutions of supporting electrolyte (0.55 M NaCl, pH ≈ 6). The above-mentioned conditions, including the supporting electrolyte selection, are chosen to be consistent and comparable with our previous measurements by the same approach [2,7,8,14,16,36]. Nevertheless, we tested the method using the electrolyte of lower ionic strength to mimic more atmospheric conditions (0.1 M NaCl) and did not detect any significant difference in the electrochemical response.

2.2. Aerosol Sampling and Preparation

Natural samples of atmospheric $PM_{2.5}$ were collected by using the low-volume Sequential Sampler SEQ 47/50 (SEQ47/50) (Sven Leckel, Ingenieuburo GmbH, Berlin, Germany) on a pre-baked (450 °C for 4 h) glass fiber filters (GF/F) (Whatman, grade GF/F, d = 47 mm). The sample collection was conducted at a flow rate of 2.3 m^3/h for 48 h.

Sampling was conducted throughout the spring and early autumn in 2016 at the Middle Adriatic location (next to marine Rogoznica Lake, 43.53° N, 15.95° E). According to the air-mass backward trajectory analysis by using the NOAA HYSPLIT model at 10 m above sea level continental air-mass inflows mostly affected sampling area in winter and autumn, while in spring and summer periods southern marine pathways had higher impact. From October to March the synoptic wind circulation was northerly, while from April to September the direction of wind changed to southerly with the domination of south western wind during the summer.

Collected aerosol samples were stored in Petri slides (Millipore Inc., Darmstadt, Germany) at −50 °C prior to analysis. Approximately 10% of the filter sample was cut in pieces and dissolved in 10–20 mL of ultra-pure water (Milli-Q, 18.2 MΩ, total organic carbon (TOC) < 3 ppb), ultrasonicated for 20 min, and stored at 4 °C overnight. Afterward, the solution was filtrated through GF/F filters of 0.7 μm pore sizes and further treated as WS aerosol fraction. For the electrochemical measurements the final ion strength of the WS solution was adjusted by adding saturated NaCl solution to correspond to 0.55 M NaCl electrolyte.

For sulfate (SO_4^{2-}) analysis approximately a quarter of each exposed filter was cut and dissolved in 10 mL of ultra-pure water (Milli-Q, 18.2 MΩ, total organic carbon (TOC) < 3 ppb), ultrasonicated for 20 min, and stored at 4 °C overnight. The solutions were filtrated through 0.22 μm PTFE disk filters and analyzed by ion chromatography (IC). SO_4^{2-} anions were measured on a Dionex ICS 3000 ion chromatograph (Thermo Scientific, Sunnyvale, CA, USA) with a conductivity detector.

2.3. Electrochemical Instrumentation and Procedure

The RSS measurements: The RSS measurements were performed with a μ-Autolab (Electrochemical Instrument Eco Chemie, Metrohm Autolab B.V., Utrecht, Netherlands) electrochemical analyzer connected to a 663 VA Stand Metrohm mercury electrode. The reference electrode was an Ag/AgCl/3 M KCl (Ag/AgCl electrode connected to the solution with an electrolyte bridge). A platinum electrode served as the auxiliary electrode. The cell volume was 25 mL.

The applied electrochemical technique was cathodic striping square wave voltammetry (CSSWV). Determination of the non-volatile RSS compounds in Hg was made by their deposition at positive electrode potentials (E_d = −0.2 and −0.4 V vs. reference electrode Ag/AgCl) before and after the acidification step and purging by N_2 to remove volatile RSS as already described [2,7,16]. Namely, non-volatile RSS can be measured by voltammetry after acidification and purging of the sample with an inert gas to remove all volatile RSS. For the given experimental conditions, purging by N_2 was performed between 3 and 5 min. Cathodic scans with and without accumulation (acc. time, t_a = 0–120 s) by stirring at the starting deposition potentials were run with an amplitude (A) of 25 mV and frequency (f) of 80 Hz. The sensitivity of the measurements with 120 s accumulation time is given by the limit of detection (LOD) and limit of quantification (LOQ), which for sulfide measurements were 0.13 and 0.46 nM, respectively, and for 3-MPA the LOD was 0.66 nM and the LOQ 2.21 nM. Quantitation of RSS in the studied samples was based on external calibration, by using calibration curves for sulfide and 3-MPA in 0.55 M NaCl.

The surface-active substance (SAS) measurements: SASs were determined by the electrochemical method of alternating current (a.c.) voltammetry with out-of-phase mode using a μ-Autolab (Electrochemical Instrument Eco Chemie, Metrohm Autolab B.V., Utrecht, Netherlands) electrochemical analyzer connected with 663 VA Stand Metrohm mercury electrode equipped as described above. A.c. voltammetry is based on the SAS adsorption effects at the hydrophobic surface of Hg (at $E_d = -0.6$ V) [2,36 and references therein]. The measurements were performed under following conditions: accumulation potential $E_d = -0.6$ V, $t_a = 30$ s, f = 75 Hz, amplitude $A = 10$ mV, and phase angle = 90°. The concentration of SAS in each sample was expressed in equivalents (mg L^{-1}) of a model surfactant Triton-X-100 (T-X-100). The detection limit of SAS determination is 0.02 mg L^{-1} equivalent of T-X-100, with LOQ of 0.05 mg L^{-1}. All electrochemical measurements were done at room temperature (21 ± 1 °C).

In addition, water-soluble organic carbon (WSOC) content was determined by the high-temperature catalytic oxidation (HTCO) method at a TOC-V_{CPH} instrument (Shimadzu, Japan) as described previously [7,16,36]. The WSOC sample aliquot (15 mL) was acidified with 2 M HCl to pH ~ 3 in order to eliminate the inorganic carbonates. The concentration of each sample was calculated as an average of three to five replicates. The quantification limit was 0.228 M for dissolved organic C with reproducibility of 5%.

3. Results and Discussion

3.1. Electrochemical Behavior of 3-MPA

During voltammetric experiments, accumulation of 3-MPA on Hg at $E_d = -0.2$ V resulted in the formation of a 3-MPA–Hg complex (Equation (2)), which was then reduced by scanning the potential toward more negative values. The reduction of the formed complex occurred at around −0.6 V by giving a reversible and reproducible reduction peak according to process explained by the reverse of Equation (2) [9,37]:

$$2RSH + Hg \leftrightarrow (RS)_2Hg_{(ads)} + 2H^+ + 2e^-. \quad (2)$$

The RSH stands here for thio-type RSS (DMS, DMDS, cysteine, methanethiol, gluthatione). In general, the position and the height of the 3-MPA–Hg reduction peak is sensitive to the pH of the bulk solution as already noted for other RSS studied at the Hg surface [2,3,12,15]. Therefore, in this study, measurements were done at the same conditions, pH ~ 6. By prolonging the accumulation time with stirring (t_a = 0–120 s) at the starting $E_d = -0.2$ V the height of the 3-MPA–Hg reduction peak increases, I_p as shown for the 9 nM 3-MPA solution (Figure 1a). The same effect is obtained by increasing the 3-MPA bulk concentrations (Figure 1b) used for the preparation of a calibration curve (Figure 1c).

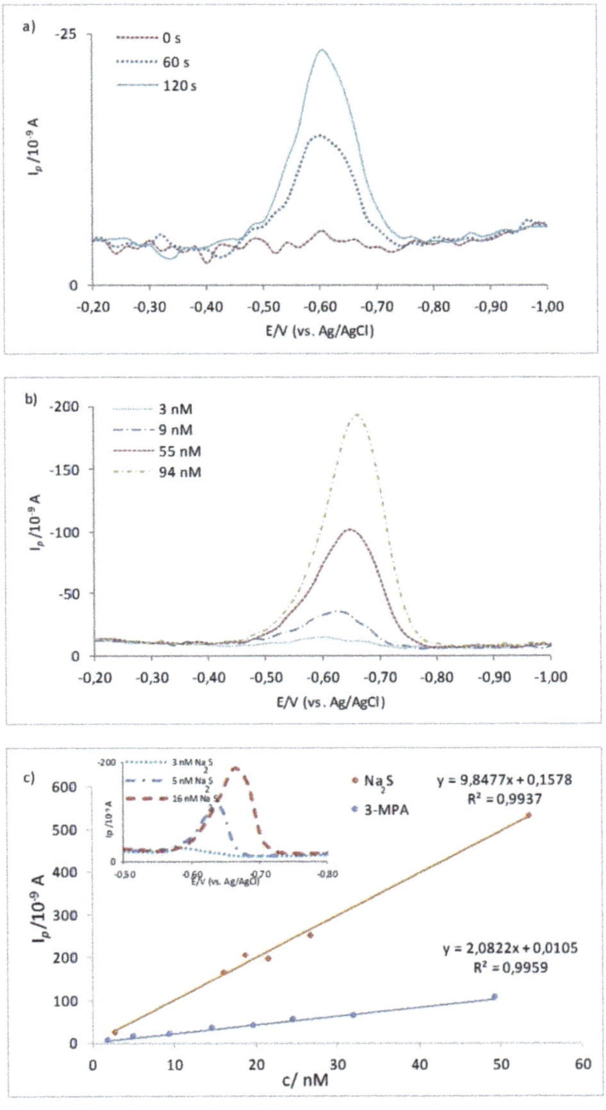

Figure 1. Cathodic striping square wave voltammetry (CSSWV) curves showing (**a**) accumulation effect in 9 nM methyl 3-mercaptopropionate (3-MPA) electrolyte solution; (**b**) effect of increasing concentration of the 3-MPA, and (**c**) calibration curves for the 3-MPA and Na_2S, with insertion of CSSWV curves for increasing sulfide concentration in electrolyte solution. Experimental conditions: $E_d = -0.2$ V, A = 25 mV, f = 80 Hz, $t_a = 120$ s.

The linear relationship between the 3-MPA concentration and I_p which is a direct measure of the reduction process at the Hg, is obtained for the wide concentration range, between 3.0 and 500 nM for an accumulation step with stirring ($t_a = 120$ s). For the given experimental conditions at concentration of 500 nM, saturation of the electrode was achieved. Considering the concentrations of RSS expected in marine aerosol samples, we focused on the calibration in the lower concentration range, between 2 and 50 nM, as presented in Figure 1c. For comparison, the calibration curve of the Na_2S is presented in the same graph. Typical voltammetric curves for increasing sulfide concentration are given as an insert in the same Figure. The height of the standard HgS reduction peak, I_p, increases sharply with increasing the concentration of Na_2S in the bulk solutions, being about four times more sensitive than the I_p of the same 3-MPA concentrations (Figure 1c), as already documented for thiols, Na_2S, and S^0 [38,39]. The peak potential of the HgS reduction, if compared with the 3-MPA-Hg reduction, is moving as well more negatively with the increase of the bulk sulfide concentration, i.e., the amount of the formed HgS layer [13,14]. Such a difference in sensitivity is used in this study as an advantage in characterization of mixtures containing organic and inorganic RSS, i.e., the 3-MPA and sulfide.

Variation of the deposition potentials (at the given experimental conditions) from E_d = 0.00 to −0.50 V has been already shown to influence electrochemistry of the organic RSS at the Hg surface [9,10,13,40], i.e., the appearance and height of the revealed reduction peaks. Such a property of setting the electrode at the unique deposition potential of maximum response for the selected model compounds was successfully used in the analytical protocol for thiol characterization in seawater and freshwater samples [6,9–11,40,41]. Depending on the deposition potential, dimethylarsinyl-ethanol sulfide [13] and similar labile compounds would be oxidized and deposit the HgS layer at the Hg surface. The same behavior is also characteristic for thiourea [42] and some other low-molecular-weight thiols (LMWTs) [9,10] which deposit HgS around 0 V. The observed behavior is caused by the different stability of the RS–Hg and HgS-type compounds as well as different lability of organic RSS at the studied Hg electrode potentials. Here, by changing the deposition potential from −0.2 toward −0.4 V, for the solution containing 3-MPA, the height of the 3-MPA–Hg reduction peak is significantly decreasing by depositing at more negative potentials (Figure 2). Sulfide and elemental sulfur (S^0) in the same range of deposition potentials would not change their electrochemical behavior [13].

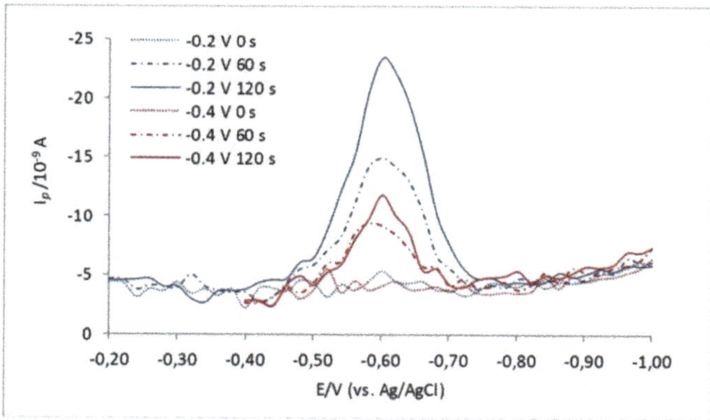

Figure 2. Effect of different deposition potentials, E_d = −0.2 V (red lines), and E_d = −0.4 V (blue lines) and different accumulation times for electrolyte solution containing 9 nM 3-MPA. Experimental conditions: A = 25 mV; f = 80 Hz.

Usually in the adopted electrochemical measurements of the RSS in natural samples, the acidification and purging step by N_2 is used to make a distinction between the volatile and non-volatile fraction of the RSS [2,3,7,13,16]. Sulfide-type RSS are considered volatile, while elemental sulfur (S^0) and S^0-like compounds (including some LMWTs that deposit HgS and are not sensitive to purging in acidic conditions) are considered non-volatile RSS that mainly contribute to the HgS peak in natural samples. The 3-MPA is shown to be quite stable in water solution. Experiments with 3-MPA approved its stability in water solution within several hours. As distinguished from sulfide which disappears completely from acidic solution (pH = 3–5) in 3–5 min depending on the concentration, the 3-MPA is insensitive to acidification and purging with nitrogen (see Figure 3). This different behavior of 3-MPA and sulfide in acidic conditions enables their direct electrochemical determination in the solution.

3.2. Mixture of Model RSS: 3-MPA and Na_2S as Representatives for Organic and Inorganic RSS

Although under some experimental conditions the RS–Hg and HgS reduction peaks merge [9,10,40], in our case at given experimental conditions: By selecting the 3-MPA as a representative of the organic RSS; by selecting the deposition potentials at E_d = −0.2 and −0.4 V, and by the different deposition accumulation times (t_a = 0−120 s), separate reduction peaks were clearly revealed as shown in Figure 3. Different accumulation times at different deposition potentials influence the position and intensities of the recorded reduction peaks in the mixture of the 3-MPA and Na_2S. Deposition at E_d = −0.2 V would produce only one peak, while deposition at E_d = −0.4 V allows appearance of both the 3-MPA–Hg and HgS reduction.

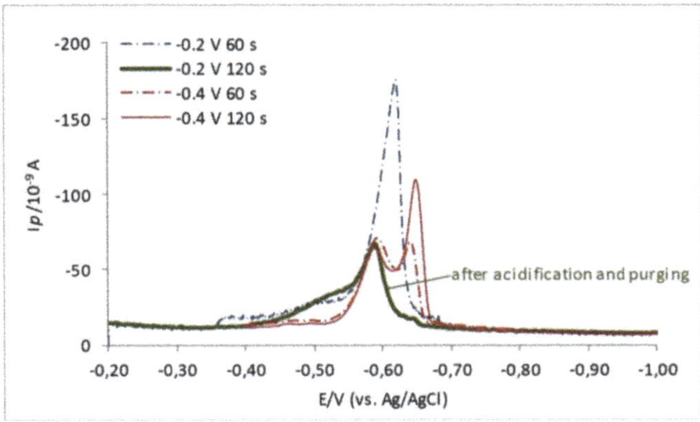

Figure 3. CSSWV curves of the solution containing 65.0 nM 3-MPA and 32.5 nM Na_2S, recorded at the different deposition potentials with different accumulation times, as illustrated. Green solid (bold) line is recorded after purging the mixture solution with N_2 when all added sulfide was stripped out. Experimental conditions: E_d = −0.2 and −0.4 V, t_a = 60 and 120 s, A = 25 mV, f = 80 Hz.

Herein, note that with longer accumulation time at E_d = −0.2, only one peak appears. While by deposition at E_d = −0.4 V, and longer deposition, the HgS reduction is also expressed. The acidification step assures that all volatile RSS (sulfide here) will be removed from the mixture, and only non-volatile RSS will remain.

However, the accumulation effect is more expressed in the case of HgS. As already shown, the HgS reduction is not influenced by changing the deposition potentials from 0 to −0.5 V [13], and the HgS reduction peak remains the same whether it is accumulated at −0.2 or −0.4 V. In other words, by selecting these potentials it is possible to make a rough distinction between organic and inorganic S compounds [2,13,16] since majority of organic RSS would not deposit HgS at Hg.

Furthermore, there is a possibility to further characterize and define recorded peaks separately by detailed convolution analyses (Autolab GPES and/or ECDSOFT software, [43]; http://gss.srce.hr/pithos/rest/omanovic@irb.hr/files/Software/.) However, if the acidification and purging step by N_2 is additionally applied, it is possible to completely remove sulfide (i.e., all volatile RSS) from the mixture and detect one reduction peak assigned to 3-MPA–Hg, as shown in Figure 3 by the green, bold line. Therefore, 3-MPA belongs to a pool of non-volatile and non-acidification-step sensitive RSS, as already discussed [2].

Based on the above-presented results, we conclude that the mixture characterization should be done by measuring at deposition potentials of −0.2 and −0.4 V, with variation of accumulation times (t_a = 0–120s). The acidification step (pH = 2–3), and the measurement of the voltammetric peaks before and after acidification and purging, followed by adjustment to original pH, will indicate the presence of volatile and/or non-volatile RSS.

3.3. Electrochemical RSS Responses in WS Fraction of Fine Marine Aerosols

Electrochemical responses recorded in the WS fraction of the fine marine aerosols are shown in Figure 4. In Figure 4a, the electrochemical response is shown for three consecutive samples from the spring season (March 29–April 4, 2016), denoted as S1, S2, and S3. Figure 4b shows electrochemical responses for four early autumn samples denoted as A1, A2, A3, and A6 (October 8–14, 2016 and October 18–20, 2016, respectively). In spring samples, the presence of mainly RS–Hg reduction was observed, while both, the RS–Hg and HgS peaks were detected in early autumn samples. Peak identification was additionally confirmed by considering the standard 3-MPA solution spiked with Na_2S as shown within Figure 4a.

Electrochemical responses recorded in the WS fraction of the fine marine aerosols are shown in Figure 4. In Figure 4a, the electrochemical response is shown for three consecutive samples from the spring season (29 March–4 April 2016), denoted as S1, S2, and S3. Figure 4b shows electrochemical responses for four early autumn samples denoted as A1, A2, A3, and A6 (October 8–14, 2016 and October 18–20, 2016, respectively). In spring samples, the presence of mainly RS–Hg reduction was observed, while both, the RS–Hg and HgS peaks were detected in early autumn samples. Peak identification was additionally confirmed by considering the standard 3-MPA solution spiked with Na_2S as shown within Figure 4a.

Observed RSS peaks in the studied WS aerosol samples were recorded in the same range of potentials (−0.58 to −0.76 V) as recorded in the model mixtures of the 3-MPA and sulfide. In spring samples, there was no difference in the peak appearance and its height by measuring at different (E_d = −0.2 and −0.4 V) deposition potentials, behaving more like the HgS reduction, even the peak position was more positive than expected for the typical HgS reduction (Figure 4a). However, in autumn samples, deposition at −0.4 V during t_a = 120 s revealed only the HgS peak (Figure 4c). In these samples, the accumulation effect was also more expressed through increasing of the HgS peak as observed in the model mixtures. Standard addition of the 3-MPA directly into the electrochemical cell containing the A1 sample after the acidification and purging step caused an increase of the first more positive peak, RS–Hg around −0.6 V (Figure 4d), pointing to similar electrochemical behavior to the 3-MPA-Hg electrode process, while the addition of sulfide caused the increase of the more negative HgS peak. It is important to stress that all recorded RSS peaks (RS–Hg and HgS-type electrode reaction) were not visible under diffusion-controlled conditions (t_a = 0s) and were not sensitive to the purging and acidification step, indicating the presence of non-volatile RSS.

The appearance of the negative HgS peak in autumn samples could be likely associated with enhanced release of volatile S compounds during water layer mixing in the nearby marine Rogoznica Lake [44]. That is, during intense vertical lake water layers mixing, millimolar concentrations of sulfide present in the anoxic bottom layer reach the surface where it is rapidly oxidized (mainly to S^0) and further lost into the atmosphere. Such a scenario is supported by decreasing the HgS peak in aerosol samples collected five days and later from October 5/6, when mixing of the water layers started (Figure 4b and Table 1). In the same WS aerosol samples, the decrease of the RS–Hg peaks was also noticed.

According to the electrochemical behavior and position of the recorded RSS peaks in the studied ambient aerosol samples, corresponding to the 3-MPA–Hg and HgS reduction, the detected peaks presumably can be evidence of non-volatile mercapto-type RSS and sulfide and/or S^0 or other non-volatile and labile RSS-like compounds that deposit HgS at the Hg surface. The existence of similar compounds has already been proved for the oxic water layers in the North Adriatic as well as marine Rogoznica Lake [2,7,13]. Similar electrochemical response, implying similar RSS were reported as well for precipitation samples (the concentration range for detected RSS was between 2 and 5 nM) [16], where mainly volatile RSS were associated with the RS–Hg peak detected at around −0.55 V (vs. Ag/AgCl).

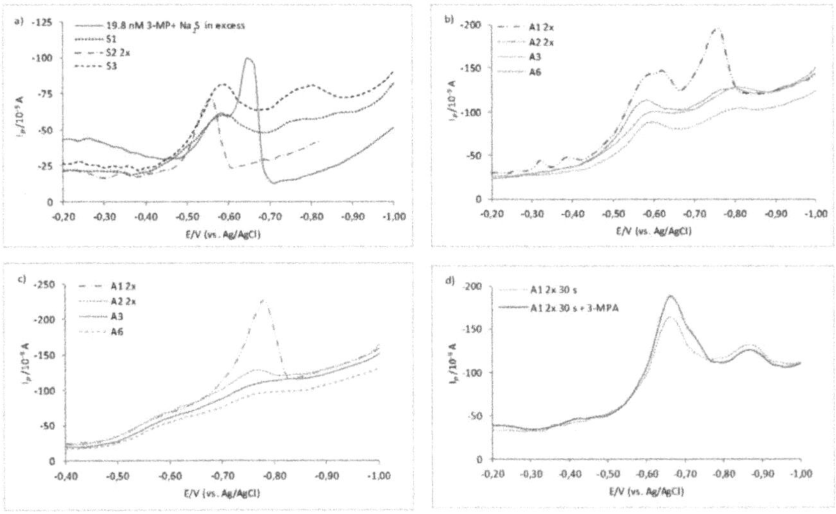

Figure 4. CSSWV curves recorded in the WS fraction of (**a**) three consecutive Spring 2016 aerosol samples (S1, S2, and S3). Red-solid line corresponds to CSSWV scan of the electrolyte containing 19.8 nM 3-MPA spiked with Na_2S; (**b**) four consecutive Autumn 2016 aerosol samples (A1, A2, A3, A6); (**c**) effect of different deposition potential (E_d = −0.4 V) for Autumn samples A1, A2, A3, A6. t_a = 120 s for **a–c**; (**d**) addition of the standard 3-MPA in Autumn sample A1, after the acidification and purging step, t_a = 30 s. Please note that pH in this sample was readjusted to 10 after acidification, which influenced the peak position. The best response in samples S2 and A1 was obtained after 2× dilution of original WS aerosol fractions. These samples have the highest surface-active substance (SAS) content. Other experimental conditions: E_d = −0.2 V, *A* = 25 mV, f = 80 Hz.

In the studied spring samples, I_p of non-volatile RS–Hg peaks ranges from 22 to 130 nA, which corresponds to concentrations of the mercapto-type compounds ranging from 10.0 to 65.0 nM in analyzed WS filter aliquots (calibrated with the 3-MPA). The concentration of non-volatile RSS, expressed in relation to the volume of the sampled air, ranges from 2.60 to 15.40 ng m^{-3} (mean value = 7.40 ng m^{-3}, *N* = 10). In autumn samples, detected peaks range from 12 to 63 nA, corresponding to concentrations of the mercapto-type compounds ranging from 6.0 to 30.0 nM, i.e., from 0.48 to 2.23 ng m^{-3} (mean value = 1.26 ng m^{-3}, *N* = 6). The I_p of the second HgS peak in autumn samples calibrated with the sulfide, ranged from 7 to 117 nA, which correspond to RSS concentration ranging from 0.75–11.89 nM in analyzed WS filter aliquots, i.e., from 0.02 to 0.26 ng m^{-3} (mean value = 0.07 ng m^{-3}, *N* = 6) if expressed in relation to the volume of the sampled air.

In addition, concentration of WSOC and surface-active substances (SAS) was determined for all aerosol samples discussed above (S1–S3 and A1–A3, A6), and their values are given in Table 1. Relatively high organic matter content, especially its surface-active fraction that strongly adsorbs on the hydrophobic Hg surface [2,36,45], could potentially suppress the reduction process of organic RSS [2,12,45] and mask the real concentration of RSS present in the ambient samples. On the other side, recorded RSS peaks in solutions with such relatively high SAS presence indicate relatively strong interaction between the present RSS and Hg. Recently published work on voltammetry of reduced glutathione, in its analytical protocol recommends a reduction of natural organic matter concentration to at least 1 mg L^{-1} [12] for getting the signal for Hg-glutathione reduction, implying that, in the samples studied here, RSS with a stronger stability constant with Hg than that with Hg-glutathione are possible present.

Presence of SAS is also known to influence the appearance and shape of the HgS peak [45], as was the case in WS aerosol samples studied here. For example, in samples S2 and A1, the most evident RSS response was obtained after double dilution of the original WS aerosol fractions. Moreover, it appears that the HgS peak is positioned more negatively in WS aerosol and precipitation samples [16] in comparison with seawater samples, which can be a consequence of more hydrophobic SAS presence that are blocking electrode redox reaction. Furthermore, the same organic material could have important role in increasing the solubility of organic compounds as well as stabilization of RSS in the aqueous phase and could modify the dissolution rate of aerosol particles in the atmosphere. It is worth noting that higher values of WSOC (0.81–3.45 μg m^{-3}) were recorded in the spring samples when higher amount of non-volatile mercapto-type RSS (2.5–15.8 ng m^{-3}) was also detected. Such seasonal difference could be attributed to biological activity, as already shown for the WSOC fraction in the marine aerosols [46]. The observed seasonal variations of the sulfate concentration shown in Table 1 also lead to the same conclusion.

Results from this work are supportive to studies reporting a significant concentration of non-sulfate sulfur species in fine aerosols [33–35,47]. Besides organosulfates, additional sulfur species such as methanesulfonates, hydroxymethanesulfonates, sulfites, sulfides, polycyclic aromatic sulfur heterocycles, and primary biological particles were reported to contribute to non-sulfate sulfur species. Depending on location and season, the likely presence of such compounds was reported for locations dominated by biogenic emissions, such as forested and agricultural regions and marine environment. As in our case, the signal for such S aerosols was most pronounced during the summer. In addition, preliminary source apportionment results indicate sulfur-containing organic aerosols as an important WSOC fraction in the same studied aerosol samples [48]. However, these S species may also have an anthropogenic source [47].

It is also important to note how the hydrophobic part of WSOC, called humic-like substances (HULIS) [49,50] as they share similar spectroscopic properties to macromolecular humic substances in terrestrial and aquatic environments [51], could have electrochemically active sulfur. That is, a high level of electrochemically active sulfur is confirmed for fulvic and humic material isolated from the lagoon and marine sediments [7,52]. In the previous paper of Frka et al. [36], a broad voltammetric peak that was recorded around a potential of −0.70 V in continental, coastal, and urban aerosol samples as well as in the isolated atmospheric HULIS, could be possibly associated with the presence of RSS. These RSS contribute to the surfactant activity, since HULIS material, i.e., HULIS concentration can be regarded as a rough upper-limit estimate of aerosol water-soluble SAS [53]. Moreover, RSS are an important part of the total S pool in aerosol samples. On the other side, HULIS could contribute to higher solubility and stabilization of RSS in the atmospheric aqueous phase, similarly to that obtained for natural organic macromolecules and elemental sulfur in seawater [7].

In addition, sulfur-containing compounds in the class of carbohydrates and proteins have been recently identified in HULIS extracts of aerosols collected in the coastal area of South Korea [54]. Moreover, protein-like components were found dominant in marine aerosols sampled over the Amundsen Sea as a result of biological activity [55]. Similar compounds are shown to be electroactive on Hg [56–58]. Likewise, a persistent behavior of the revealed voltammetric peaks, suggest that, in this study, detected RSS are from secondary processing in the atmosphere and not from primary emissions.

Table 1. Concentrations of reduced sulfur species (RSS), water-soluble organic carbon (WSOC), surface-active substances (SAS), and sulfate (SO_4^{2-}) in the selected Spring and Autumn 2016 aerosol samples, whose electrochemical characterization is shown in Figure 4.

Aerosol Sample	Sampling Period	WSOC (μg m^{-3})	SAS eq. T-X-100 (μg m^{-3})	RSS (ng m^{-3}) Mercapto Type	RSS (ng m^{-3}) Inorganic Type	SO_4^{2-} (μg m^{-3})
S1	29.03–31.03.2016.	2.36	0.18	2.62	-	3.00
S2	31.03–02.04.2016.	2.22	0.30	9.42	-	2.33
S3	02.04–04.04.2016.	2.20	0.23	3.68	-	4.22
A1	08.10–10.08.2016.	1.84	0.23	2.23	0.26	2.03
A2	10.10–12.08.2016.	1.18	0.21	1.46	0.05	1.58
A3	12.10–14.10.2016.	0.97	0.21	1.00	0.04	1.77
A6	18.10–20.10.2016.	1.34	0.23	1.02	0.03	1.50

4. Conclusions

In this work, voltammetric methods have been used for the first time to study the presence of reduced sulfur species (RSS) in the water-soluble fraction of the fine marine aerosols collected during spring and early autumn in the area of Rogoznica Lake, a unique anoxic and sulfidic seawater environment at the Adriatic. Two types of non-volatile RSS peaks were detected and attributed to the presence of mercapto-type (RS–Hg peak at around −0.6 V and positive, calibrated with 3-MPA) and sulfide/S^0-like compounds (HgS peak more negative than −0.6 V, calibrated with sulfide). In spring samples only, the presence of the mercapto-type compounds was evident and proved by the addition of the standard 3-MPA, ranging from 2.60 to 15.40 ng m^{-3}, while both, the mercapto-type (0.48–2.23 ng m^{-3}) and sulfide and/or S^0-like compounds (0.02–0.26 ng m^{-3}) were detected in early autumn samples collected during anoxic holomictic event [44] in the nearby marine lake water column.

Intensity of the RS–Hg peaks in ambient samples varied through the studied seasons with more expressed and defined peaks recorded in the spring, which possibly indicates their association with biological activity. It is worth to note that these mercapto-type RSS, show the same seasonal trend as biogenically derived sulfate and methanesulfonic acid (MSA) recorded in the same aerosol samples [59], which as the oxidation product of dimethyl sulfide, is widely used as a proxy for marine biogenic sources [32,46,59]. MSA was shown not to be electrochemically active at the given experimental conditions.

In this preliminary work, the 3-MPA is used as a representative of LMW reduced organosulfur compounds that can be usually found in natural aqueous environment, as detected by chromatographic methods [60–62]. Thus, we do not intend to declare exactly the presence of 3-MPA in the studied samples, but the presence of similar RSS organic compounds should be considered with regard to the electrochemical response. This statement is supported by preliminary source apportionment results indicating sulfur-containing organic aerosols as an important WSOC fraction in the same aerosol samples [48]. In addition, higher values of WSOC (0.81–3.45 µg m^{-3}) were recorded in the spring samples when higher amounts of non-volatile mercapto-type RSS (2.60–15.40 ng m^{-3}) were also detected.

The adopted protocol by selecting appropriate experimental conditions (accumulation at deposition potentials E_d = −0.2 V and/or −0.4 V vs. Ag/AgCl and the acidification and purging step followed by adjustment of the pH to the original after purging) confirms that electrochemistry is a fast, direct, and non-destructive methodology for the analyses of different seawater and freshwater samples [2,6,9–11,40,41,45] and can be successfully applied for the single-scan characterization of the electrochemically active sulfur compounds in ambient aerosol samples. Further work is planned to test other atmospheric RSS-relevant substances and their possible contribution to the detected RSS peaks with possible optimization of the experimental conditions given here.

Author Contributions: Methodology, including sampling and filter sample preparation and measurements I.C. and S.F.; resources, funding: I.C.; measurements, analysis: A.C.K.; supervision: I.C. and S.F.; writing—original draft, and revised version, A.C., and I.C.

Funding: This research was funded by Hrvatska Zaklada za Znanost, grant numbers IP-11-2013-1205 SPHERE, IP-01-2018-1717 MARRES.

Acknowledgments: This work is supported by the Croatian Science Foundation projects IP-11-2013-1205 SPHERE and IP-01-2018-1717 MARRES. Authors thank Ana Kroflič and Irena Grgić from the Department of Analytical Chemistry, National Institute of Chemistry, Ljubljana, Slovenia for sulfate analysis"D. Omanović for help in using ECDSOFT software during analysis of the RSS peaks, and the anonymous reviewers for their valuable comments that improved the manuscript. Marina Frapa Rogoznica is highly acknowledged for support during aerosol sampling.

Conflicts of Interest: The authors declare that they have no conflict of interest.

Abbreviations

methyl 3-mercaptopropionate (3-MPA).

References

1. Luther, G.W.; Tsamakis, E. Concentration and form of dissolved sulfide in the oxic water column of the ocean. *Mar. Chem.* **1989**, *27*, 165–177. [CrossRef]
2. Ciglenečki, I.; Ćosović, B. Electrochemical Study of Sulfur Species in Seawater and Marine Phytoplankton Cultures. *Mar. Chem.* **1996**, *52*, 87–97. [CrossRef]
3. Ciglenečki, I.; Ćosović, B. Electrochemical Determination of Thiosulfate in Seawater in the Presence of Elemental Sulfur and Sulfide. *Electroanalysis* **1997**, *9*, 775–780. [CrossRef]
4. Al-Farawati, R.K.; Van den Berg, C.M.G. The Determination of Sulfide in Seawater by Flow-Analysis with Voltammetric Detection. *Mar. Chem.* **1997**, *57*, 277–286. [CrossRef]
5. Rozan, T.F.; Theberge, S.M.; Luther, G. Quantifying elemental sulfur (S0), bisulfide (HS-) and polysulfides (Sx2-) using a voltametric method. *Anal. Chim. Acta* **2000**, *415*, 175–184. [CrossRef]
6. Laglera, L.M.; Van den Berg, C.M.G. Copper complexation by thiol compounds in estuarine waters. *Mar. Chem.* **2003**, *82*, 71–89. [CrossRef]
7. Bura-Nakić, E.; Helz, G.R.; Ciglenečki, I.; Ćosović, B. Seasonal variations in reduced sulfur species in a stratified seawaterlake (Rogoznica Lake, Croatia): Evidence for organic carriers of reactive sulfur. *Geochim. Cosmochim. Acta* **2009**, *73*, 3738–3751. [CrossRef]
8. Bura-Nakić, E.; Viollier, B.; Ciglenečki, I. Electrochemical and colorimetric measurements show the dominant role of FeS in a permanently anoxic lake. *Environ. Sci. Technol.* **2013**, *47*, 741–749. [CrossRef]
9. Superville, P.-J.; Pižeta, I.; Omanović, D.; Billon, G. Identification and on-line monitoring of reduced sulphur species (RSS) by voltammetry in oxic waters. *Talanta* **2013**, *112*, 55–62. [CrossRef]
10. Laglera, L.M.; Downes, J.; Tovar-Sánchez, A.; Monticelli, D. Cathodic pseudopolarography: A new tool for the identification and quantification of cysteine, cystine and other low molecular weight thiols in seawater. *Anal. Chim. Acta* **2014**, *836*, 24–33. [CrossRef]
11. Marie, L.; Pernet-Coudrier, B.; Waeles, M.; Gabon, M.; Riso, R. Dynamics and sources of reduced sulfur, humic substances and dissolved organic carbon in a temperate river system affected by agricultural practices. *Sci. Total Environ.* **2015**, *537*, 23–32. [CrossRef] [PubMed]
12. Chen, W.; Guéguen, C.; Smith, D.S. Assessing effects of pH, metal ion and natural organic matter on identification and determination of reduced glutathione by cathodic stripping voltammetry. *Int. J. Environ. Anal. Chem.* **2017**, *97*, 330–334. [CrossRef]
13. Krznarić, D.; Ciglenečki, I.; Ćosović, B. Voltammetric investigations of 2-dimethylarsinyl-ethanol sulphide in NaCl and seawater. *Anal. Chim. Acta* **2001**, *431*, 269–278. [CrossRef]
14. Milanović, I.; Krznarić, D.; Bura-Nakić, E.; Ciglenečki, I. Deposition and dissolution of metal sulfide layers at a Hg electrode surface in seawater electrolyte conditions. *Environ. Chem.* **2013**, *11*, 167–172. [CrossRef]
15. Florence, T.M. Cathodic stripping voltammetry: Part I. Determination of organic sulfur compounds, flavins and porphyrins at the sub-micromolar level. *J. Electroanal. Chem.* **1979**, *97*, 219–236. [CrossRef]
16. Orlović-Leko, P.; Omanović, D.; Ciglenečki, I.; Vidović, K.; Brenko, T. Application of electrochemical methods in the physico-chemical characterization of atmospheric precipitation. *Bulg. Chem. Commun.* **2017**, *49*, 211–217.
17. Williams, K.D.; Jones, A.; Roberts, D.L.; Senior, C.A.; Woodage, M.J. The response of the climate system to the indirect effects of anthropogenic sulfate aerosol. *Clim. Dyn.* **2001**, *17*, 845–856. [CrossRef]
18. Burnett, R.T.; Dales, R.; Krewski, D.; Vincent, R.; Dann, T.; Brook, J.R. Associations between ambient particulate sulfate and admissions to Ontario hospitals for cardiac and respiratory diseases. *Am. J. Epidemiol.* **1995**, *142*, 15–22. [CrossRef]
19. Pardo, M.; Porat, Z.; Rudich, A.; Schauer, J.J.; Rudich, Y. Repeated exposures to roadside particulate matter extracts suppresses pulmonary defense mechanisms, resulting in lipid and protein oxidative damage. *Environ. Pollut.* **2016**, *210*, 227–237. [CrossRef]
20. Galloway, J.N. Acid deposition: Perspectives in time and space. *Water Air Soil Pollut.* **1995**, *85*, 15–24. [CrossRef]
21. Charlson, R.J.; Lovelock, J.E.; Andreae, M.O.; Warren, S.G. Oceanic phytoplankton, atmospheric sulphur, cloud albedo and climate. *Nature* **1987**, *326*, 655–661. [CrossRef]
22. Quinn, P.K.; Bates, T.S. The case against climate regulation via oceanic phytoplankton sulphur emissions. *Nature* **2011**, *480*, 51–56. [CrossRef] [PubMed]

23. Cullis, C.F.; Hirschler, M.M. Atmospheric sulfur: Natural and man-made sources. *Atmos. Environ.* **1980**, *14*, 1263–1278. [CrossRef]
24. Chin, M.; Jacob, D.J.; Gardner, G.M.; Foreman-Fowler, M.S.; Spiro, P.A.; Savoie, D.L. A global three-dimensional model of tropospheric sulfate. *J. Geophys. Res.* **1996**, *101*, 18667–18690. [CrossRef]
25. Smith, S.J.; Pitcher, H.; Wigley, T.M.L. Global and Regional Anthropogenic Sulfur Dioxide Emissions. *Glob. Planet. Change* **2001**, *29*, 99–119. [CrossRef]
26. Graf, H.F.; Langmann, B.; Feichter, J. The contribution of Earth degassing to the atmospheric sulfur budget. *Chem. Geol.* **1998**, *147*, 131–145. [CrossRef]
27. Bhugwant, C.; Siéja, B.; Bessafi, M.; Staudacher, T.; Ecormier, J. Atmospheric sulfur dioxide measurements during the 2005 and 2007 eruptions of the Piton de La Fournaise volcano: Implications for human health and environmental changes. *J. Volcanol. Geotherm. Res.* **2009**, *184*, 208–224. [CrossRef]
28. Kristmannsdottir, H.; Sigurgeirsson, M.; Armannsson, H.; Hjartarson, H.; Olafsson, M. Sulfur gas emissions from geothermal power plants in Iceland. *Geothermics* **2000**, *29*, 525–538. [CrossRef]
29. Andreae, M.O. Ocean-atmosphere Interactions in the global biogeochemical sulfur cycle. *Mar. Chem.* **1990**, *30*, 1–29. [CrossRef]
30. Malin, G.; Kirst, G.O. Algal production of dimethyl sulfide and its atmospheric role. *J. Phycol.* **1997**, *33*, 889–896. [CrossRef]
31. Stefels, J.; Steinke, M.; Turner, S.; Malin, G.; Belviso, S. Environmental constraints on the production and removal of the climatically active gas dimethylsulphide (DMS) and implications for ecosystem modeling. *Biogeochemistry* **2007**, *83*, 245–275. [CrossRef]
32. Seinfeld, J.H.; Pandis, S.N. *Atmospheric Chemistry and Physics from Air Pollution to Climate Change*, 2nd ed.; John Wiley & Sons: New York, NY, USA, 2006.
33. Neubauer, K.R.; Sum, S.T.; Johnston, M.V.; Wexler, A.S. Sulfur speciation in individual aerosol particles. *J. Geophys. Res.* **1996**, *101*, 18701–18707. [CrossRef]
34. Graham, B.; Guyon, P.; Maenhaut, W.; Taylor, P.E.; Ebert, M.; Matthias-Maser, S.; Mayol-Bracero, O.L.; Godoi, R.H.M.; Artaxo, P.; Meixner, F.X.; et al. Composition and diurnal variability of the natural Amazonian aerosol. *J. Geophys. Res.* **2003**, *108*, 4765–4782. [CrossRef]
35. Cozzi, F.; Pellergrini, I.; Adami, G.; Reisenhofer, E.; Bovenzi, M.; Barbieri, P. Sulphur speciation of PM10 samples by XANES spectroscopy. *Cent. Eur. J. Chem.* **2009**, *7*, 395–401. [CrossRef]
36. Frka, S.; Dautović, J.; Kozarac, Z.; Ćosović, B.; Hoffer, A.; Kiss, G. Surface-active substances in atmospheric aerosol: An electrochemical. *Tellus B Chem. Phys. Meteorol.* **2012**, *64*, 18490–18503. [CrossRef]
37. Luther, G.W.; Church, T.M.; Giblin, A.E.; Howarth, R.W. Speciation of Dissolved Sulfur in Salt Marshes by Polarographic Methods. In *Organic Marine Geochemistry*; Symposium Series; Mary, L.J., Ed.; Marine Chemistry in the Coastal Environment, American Chemical Society: Washington, DC, USA; Volume 305, pp. 340–355, Chapter 20.
38. Renard, J.J.; Kubes, G.; Bolker, H.I. Polarographic determination of sulfur compounds in pulping liquors. *Anal. Chem.* **1975**, *47*, 1347–1352. [CrossRef]
39. Wang, F.; Tessier, A. Voltammetric determination of elemental sulfur in pore waters. *Limnol. Oceanogr.* **1998**, *43*, 1353–1361. [CrossRef]
40. Laglera, L.M.; Tovar-Sanchez, A. Direct recognition and quantification by voltammetry of thiol/ thioamide mixes in seawater. *Talanta* **2012**, *89*, 496–504. [CrossRef]
41. Pernet-Coudrier, B.; Waeles, M.; Filella, M.; Quentel, F.; Riso, R.D. Simple and simultaneous determination of glutathione, thioacetamide and refractory organic matter in natural waters by DP-CSV. *Sci. Total Environ.* **2013**, *463*, 997–1005. [CrossRef]
42. Banica, F.-G.; Galik, M.; Švancar, I.; Vytras, K. Electrochemical Investigation of Metal Sulfides at Mercury Electrodes Using Thiourea as a Source of Sulfide Ion. *Electroanalysis* **2009**, *21*, 332–341. [CrossRef]
43. Omanović, D.; Branica, M. Automation of voltammetric measurements by polarographic analyser PAR 384B. *Croat. Chem. Acta* **1998**, *71*, 421–433.
44. Ciglenečki, I.; Ljubešić, Z.; Janeković, I.; Batistić, M. *Rogoznica Lake, A Euxinic Marine Lake on the ADRIATIC Coast (Croatia) that Fluctuates Between Anoxic Holomictic and Meromictic Conditions*; Ramesh, D.G., Zadereev, E.S., Degermendzhi, A.G., Eds.; Springer: Wageningen, The Netherlands, 2017; pp. 125–154.
45. Ciglenečki, I.; Kodba, Z.; Ćosović, B. Sulfur Species in Rogoznica Lake. *Mar. Chem.* **1996**, *53*, 101–110. [CrossRef]

46. Yoon, Y.J.; Ceburnis, D.; Cavalli, F.; Jourdan, O.; Putaud, J.P.; Facchini, M.C.; Decesari, S.; Fuzzi, S.; Sellegri, K.; Jennings, S.G.; et al. Seasonal characteristics of the physicochemical properties of North Atlantic marine atmospheric aerosols. *J. Geophys. Res. Atoms. Geophys. Union* **2007**, *112*, D04206. [CrossRef]
47. Shakya, K.M.; Peltier, R.E. Non-sulfate sulfur in fine aerosols across the United States: Insight for organosulfate prevalence. *Atmos. Environ.* **2015**, *100*, 159–166. [CrossRef] [PubMed]
48. Casotto, R.; Cvitešić Kušan, A.; Bhattu, D.; Ciglenečki, I.; Frka, S.; Kroflič, A.; Grgić, I.; Baltensperger, U.; Slowik, J.; Prévôt, A.S.H. Combined analysis using AMS and EESI measures for organic aerosol source apportionment of the Adriatic coast, poster presentation (P1-054). In Proceedings of the European Aerosol Conference, Gothenburg, Sweden, 25–30 August 2019.
49. Kiss, G.; Tombacz, E.; Hansson, H.C. Surface tension effects of humic-like substances in the aqueous extract of tropospheric fine aerosol. *J. Atmos. Chem.* **2005**, *50*, 279–294. [CrossRef]
50. Facchini, M.C.; Decesari, S.; Mircea, M.; Fuzzi, S.; Loglio, G. Surface tension of atmospheric wet aerosol and cloud/fog droplets in relation to their organic carbon content and chemical composition. *Atmos. Environ.* **2000**, *34*, 4853–4857. [CrossRef]
51. Kiss, G.; Varga, B.; Galambos, I.; Ganszky, I. Characterization of water-soluble organic matter isolated from atmospheric fine aerosol. *J. Geophys. Res. Atmos.* **2002**, *107*, D21. [CrossRef]
52. Gašparović, B.; Ćosović, B. Electrochemical estimation of the dominant type of surface active substances in seawater samples using o-nitrophenol as a probe. *Mar. Chem.* **1994**, *46*, 179–188. [CrossRef]
53. Kroflič, A.; Frka, S.; Simmel, M.; Wex, H.; Grgić, I. Size-resolved surface active substances of atmospheric aerosol: Reconsideration of 3 the impact on sloud droplet formation. *Environ. Sci. Technol.* **2018**, *52*, 9179–9187. [CrossRef]
54. Jang, K.-S.; Choi, A.Y.; Choi, M.; Kang, H.; Kim, T.-W.K.; Park, K.T. Size-Segregated chemical compositions of HULISs in ambient aerosols collected during the winter season in Songdo, South Korea. *Atmosphere* **2019**, *10*, 226–240. [CrossRef]
55. Jung, J.; Hong, S.-B.; Chen, M.; Hur, J.; Jiao, L.; Lee, Y.; Park, K.; Hahm, D.; Choi, Y.-O.; Yang, E.J.; et al. Characteristics of biogenically-derived aerosols over the Amundsen Sea, Antarctica. *Atmos. Chem. Phys. Discuss.* **2019**, *3*. [CrossRef]
56. Ciglenečki, I.; Plavšić, M.; Vojvodić, V.; Ćosović, B.; Pepi, M.; Baldi, F. Mucopolysaccharide transformation by sulfide in diatom culture and natural mucilage. *Mar. Ecol. Prog. Ser.* **2003**, *263*, 17–27. [CrossRef]
57. Strmečki, S.; Plavšić, M.; Steigenberger, S.; Passow, U. Characterization of phytoplankton exudates and carbohydrates in relation to their complexation of copper, cadmium and iron. *Mar. Ecol. Prog. Ser.* **2010**, *408*, 33–46. [CrossRef]
58. Strmečki, S.; Paleček, E. Adsorption/desorption of biomacromolecules involved in catalytic hydrogen evolution. *Bioelectrochemistry* **2018**, *120*, 87–93. [CrossRef] [PubMed]
59. Cvitešić Kušan, A.; Kroflič, A.; Grgić, I.; Ciglenečki, I.; Frka, S. Chemical characterization of fine aerosols in respect to water-soluble ions at the eastern Middle Adriatic coast. *Environ. Sci. Pollut. Res.* **2019**. Submitted to.
60. Shea, D.; MaCcre, W.A. Determination of hydrophilic thiols in sediment porewater using ion-pair liquid chromatography coupled with electrochemical detection. *Anal. Chem.* **1988**, *60*, 1449–1454. [CrossRef]
61. Saltzman, E.S.; Cooper, W.J. *Biogenic Sulfur in the Environment*; American Chemical Society: Washington, DC, USA, 1989; Volume 39.
62. Eitel, E.M.; Taillefert, M. Mechanistic investigation of Fe(III) oxide reduction by low molecular weight organic sulfur species. *Geochim. Cosmochim. Acta* **2017**, *215*, 173–188. [CrossRef]

atmosphere

MDPI

Article

Aqueous-Phase Production of Secondary Organic Aerosols from Oxidation of Dibenzothiophene (DBT)

Yu Liu [1], Junchen Lu [1], Yanfang Chen [1], Yue Liu [1], Zhaolian Ye [2] and Xinlei Ge [1,*]

[1] Jiangsu Key Laboratory of Atmospheric Environment Monitoring and Pollution Control, Collaborative Innovation Center of Atmospheric Environment and Equipment Technology, School of Environmental Sciences and Engineering, Nanjing University of Information Science and Technology, Nanjing 210044, China; 20171207360@nuist.edu.cn (Y.L.); a15189821766@163.com (J.L.); chenyf920121@163.com (Y.C.); liuyue1019ly@163.com (Y.L.)

[2] College of Chemistry and Environmental Engineering, Jiangsu University of Technology, Changzhou 213001, China; bess_ye@jsut.edu.cn

* Correspondence: caxinra@163.com; Tel.: +86-25-58731394

Received: 13 January 2020; Accepted: 29 January 2020; Published: 30 January 2020

Abstract: Intermediate-volatility organic compounds (IVOCs) have been recognized as an important contributor to the secondary organic aerosol (SOA) formation via gas-phase reactions. However, it is unclear whether or not IVOCs-SOA can be produced in the aqueous phase. This work investigated aqueous oxidation of one model compound of IVOCs, dibenzothiophene (DBT). Results show that DBT can be degraded by both hydroxyl radical and the triplet excited states of organic light chromophores ($^3C^*$). Aqueous dark oxidation of DBT was also possible. SOA yields of 32% and 15% were found for hydroxyl radical (OH)-mediated photo-oxidation and dark oxidation, respectively. A continuous and significant increase of oxidation degree of SOA was observed during OH photo-oxidation, but not during the dark oxidation. Factor analyses revealed that there was a persistent production of highly oxygenated compounds from the less oxygenated species. OH-initiated photochemical reactions can also produce species with a relatively large light-absorbing ability, while such photo-enhancement due to direct light irradiation and $^3C^*$-initiated oxidation could occur, but is much less important. In the future, studies on the second-order rate constants, molecular characterization of the oxidation products from this and other IVOCs precursors are needed to better understand the role of this reaction pathway in SOA budget, air quality and climate change.

Keywords: dibenzothiophene; intermediate-volatility organic compounds; secondary organic aerosol; aqueous phase; brown carbon

1. Introduction

Secondary organic species or aerosols (SOA) are important yet highly complicated constituents of atmospheric particles, and their formation mechanism is much less clear compared to their inorganic companions [1,2]. Traditionally, SOA is thought to be generated mostly via gas-phase oxidation reactions, but in recent decades, it is well established that SOA can also be produced in atmospheric aqueous phases (e.g., fog and cloud droplets, aqueous aerosols) effectively [3,4]. This SOA from aqueous-phase reactions, termed as "aqSOA", may contribute as much mass as the gas-phase reactions [3]. On the other hand, intermediate-volatility organic compounds (IVOCs) with effective saturation concentrations (C*) in the range of 10^3–10^6 $\mu g\ m^{-3}$ [5] are proved to be an important group of precursors of SOA. These species have been observed in vehicle [6,7], ship [8], aircraft emissions [9], as well as biogenic emissions [10]. Most studies regarding SOA formation from IVOCs, however, focus on the gas-phase reactions and production yields [11–13], while the possibility and importance of IVOCs-aqSOA remain unclear. This is largely due to that IVOCs may have low water solubility or Henry's law coefficient, making it an unlikely effective aqSOA precursor in a theoretical view. Nevertheless, some phenolic carbonyls [14] or polycyclic aromatic hydrocarbon (PAH) compounds [15], which actually belong to the IVOCs category, are found to be effective in producing aqSOA. Very recently, our group also shows that 4-ethylphenol (with a C* in the high-end of IVOCs) [16], naphthoquinone and phenanthrene (both with C* values in the low-end of IVOCs) [17] all have significant aqSOA yields. These findings call more investigations on a wider coverage of IVOCs species to understand its role in the SOA budget.

In this work, we chose dibenzothiophene (DBT) as one example compound to investigate the possible role of aqueous processing on IVOCs. DBT is one of the PAH compounds, and has been detected in ambient air [10]. DBT is also linked with crude oil and petroleum from which it is removed before further use [18,19], as well as wood-burning [20]. In some cases, DBT is of interest in monitoring marine biota, water and sediments [21]. DBT has a vapor pressure of 2.05×10^{-4} mm Hg [22] and a C* of ~2.2×10^3 μg m^{-3}, falling into the low-end of IVOCs. The goal of this work is to study the degradation of DBT, production yields, chemical and optical properties of SOA derived from aqueous oxidation of DBT, highlighting the probability of IVOCs-aqSOA formation and providing parameters likely useful for atmospheric model simulations.

2. Experiments

2.1. Aqueous Oxidation Experiments

In this work, we investigated aqueous oxidation of DBT (99%, Shanghai Macklin) with two types of oxidants. One is hydroxyl radical (OH), introduced by the photolysis of hydrogen peroxide (H_2O_2, 30%, Sinopharm Chemical Reagent). Another oxidant is organic triplet excited state ($^3C^*$), which is a specific oxidant that only exists in condensed phases and is generated via light absorption by organic chromophores [23]. In this work, we used 3,4-dimethoxybenzaldehyde (DMB) (99%, Sigma-Aldrich) as the photosensitizer to generate $^3C^*$. The reaction solutions were 10 μM DBT with 300 μM H_2O_2 or with 15 μM DMB. A 10 μM amount is higher than the real concentration of DBT in ambient air, and the choice of this concentration is mainly due to the SP-AMS measurement limitations. DBT-aqSOA characterization would be heavily influenced by the background organic impurities in purified water if a lower concentration is used, while if we use a higher concentration, it may be not dissolved completely. A high H_2O_2 concentration was used to speed up the oxidation (in the order of hours), while a low concentration of DMB (relative to H_2O_2 concentration) was to make the degradation rate by $^3C^*$ comparable to that by OH. In addition, 10.0 mg L^{-1} ammonium sulfate was added into the reaction solution as an internal standard [24–26]. Sulfuric acid (96%, Sinopharm Chemical Reagent) was used to adjust the solution pH to ~5 (close to real values in fog and cloud waters [27]). The pH of the starting solution was checked with a pH meter before each experiment.

Oxidation experiments were conducted in a photo-reactor (BILON, Shanghai, China) equipped with a 1000 W Xe lamp as the light source [17]. The Xe lamp emits a broad UV–Vis light spectrum and those with wavelengths <290 nm were filtered out by Pyrex glass to mimic tropospheric sunlit conditions. Six 150 mL glass vials, each containing 100 mL of solution, were placed equidistantly around the lamp in a circular manner. The temperature of the reaction vessel was controlled at ~25 °C during measurements. The vials were well sealed; therefore, evaporation of precursor/product was expected to be negligible. The vials were wrapped in alumina foils and lights were turned off during dark aging experiments.

2.2. Kinetic Analysis

The concentration of DBT in each sampled solution during oxidation was immediately determined by using a high-performance liquid chromatograph (HPLC) (Waters, USA) with a UV–Vis detector (wavelength of 254 nm was selected) and a C18-WP column (4.6×250 mm, 5 μm). The mobile phase was a mixture of methanol and water (w/w, 9:1; isocratic method) with a flow rate of 1 mL/min. The decay of DBT typically obeys apparent pseudo-first-order kinetics, therefore follows the equation below:

$$\frac{(\mathrm{DBT})_t}{(\mathrm{DBT})_0} = \exp(-kt) \tag{1}$$

Here, $(\mathrm{DBT})_0$ and $(\mathrm{DBT})_t$are the initial concentrations of DBT (time = *0*) and that during oxidation (time = *t*). The rate constant (*k*) can be obtained by fitting the experimental data to Equation (1). Half-lifetimes of DBT oxidations ($t_{1/2}$) under different scenarios can then be estimated.

2.3. Optical Analysis

Organic carbon (OC) contents of reacted solutions were determined by a total organic carbon (TOC) analyzer (TOC-LCPH, Shimadzu, Japan), and the light absorption spectra were then analyzed by an ultraviolet–visible (UV–Vis) spectrophotometer (8453, Agilent, USA). The light absorption can be converted into an absorption coefficient at a wavelength λ (Abs_λ) [28,29] according to Equation (2). A_λ refers to absorbance at wavelength λ, A_{700} is used to correct the baseline shift (average value of 695 nm–705 nm) and *L* is the optical path length of the quartz tube (1 cm). Furthermore, we calculated the mass absorption efficiency at 365 nm (MAE_{365}, in $m^2\ g^{-1}$ C), as an indicator of the light-absorbing ability of the organic species ("brown carbon" (BrC)) generated during aqueous oxidation. The calculation of MAE_{365} is described in Equation (3). Here, C_{mass} is the OC concentration measured by the TOC analyzer.

$$\mathrm{Abs}_\lambda = (A_\lambda - A_{700}) \cdot \frac{\ln(10)}{L} \tag{2}$$

$$\mathrm{MAE}_{365} = \frac{\mathrm{Abs}_{365}}{C_{mass}} \tag{3}$$

2.4. Chemical and Factor Analysis

Another series of OH-mediated experiments (as well as the dark control experiment) was conducted by connecting the photo-reactor with an Aerodyne soot particle aerosol mass spectrometer (SP-AMS) [30,31]. The SP-AMS enables us to monitor changes of aqSOA bulk composition in the form of 70 eV electron impact (EI) ionized mass spectrum, and concentration in near real-time (1.5 min time resolution in this work). Experimental setup, SP-AMS operation and data analysis were similar to those described previously in our work [17,32]. Briefly, the reaction vial was connected to a thin line to continuously deliver solution into a clean vial for SP-AMS measurements, and another line allowing the solution to flow back. The reacted solution was then nebulized by using an atomizer. The mist was dehumidified by a diffusion dryer, and the remaining particles were sent to the SP-AMS for analyses. The nebulization mimicked the evaporation of cloud/fog droplets in real atmosphere. Non-refractory species in the particles were evaporated at 600 °C, ionized by 70 eV EI and their mass spectra were recorded by the SP-AMS. The AMS data was processed by using the Igor-based software toolkit SQUIRREL (version 1.56D) and PIKA (version 1.15D).

The mass yields of aqSOA (Y_{SOA}) were estimated by using Equation (4) below:

$$Y_{SOA} = \frac{(\mathrm{aqOrg})_{t,AMS} - (\mathrm{aqOrg})_{0,AMS}}{(\mathrm{Precursor})_{reacted}} \tag{4}$$

Here, $(\mathrm{aqOrg})_{t,\,AMS}$ and $(\mathrm{aqOrg})_{0,\,AMS}$ refer to the concentrations of organic matter detected by AMS (converted to mg L^{-1} with reference to known sulfate concentration) at time *t* and time *0*, respectively. $(\mathrm{Precursor})_{reacted}$ is the amount of consumed precursor, which can be estimated by the rate expressions determined in Section 2.2.

Moreover, we employed the positive matrix factorization (PMF) model [33] to deconvolve the mass spectral matrix of aqOrg measured by the SP-AMS. The PMF analyses aim to resolve a few factors that can reconstruct the observed time series and mass spectra of aqOrg. The resolved factor is a collection of species with similar properties rather than a single component. Here, ions with mass-to-charge ratio (*m/z*) up to 300 amu (atomic mass unit) were used for PMF analyses. Note the chemical resolution of SP-AMS is limited to identify ions with *m/z* >150, therefore we only fitted those with significant signal-to-noise (S/N) ratios. The directly measured organic data matrices, as well as the error matrices, were normalized by sulfate concentrations. The organic matrices were pre-processed according to the procedure listed by Zhang et al. [34,35], including downweight of "weak" values (0.2 < S/N < 2), removal of "bad" values (S/N < 0.2) and downweight of outliers (specific *m/z*s or time points that lead to large biases on PMF results). We explored the PMF solutions with different numbers of factors and by varying the f_{peak} from −1 to 1 with a step of 0.1 for 1 to 4 factors. No significant differences were found for the same solution at different f_{peak} values, we therefore chose the one at f_{peak} of 0. We finally chose a 3-factor solution to be the best one, and a 3-factor solution is actually common for a few prior PMF analyses on the chamber or aqueous-phase-derived SOA [14,18,36].

3. Results and Discussion

3.1. Precursor Degradation and Mass Yields of aqSOA

Concentrations of DBT during aqueous oxidations under different conditions are shown in Figure 1. Based on Figure 1, we calculated the pseudo-first-order rate constants (and uncertainties) and half-lives, as shown in Table 1. The OH-mediated photochemical degradation of DBT appeared to be the fastest, with a ~60% decay in 420 min, corresponding to a rate constant of $2.03*10^{-3}$ min^{-1} and a half-life of 342 min. Photo-oxidation by $^3C^*$ was also significant, but the rate was slower (~32% degradation in 420 min, with a rate constant of 2.03×10^{-3} min^{-1} and half-life of 668 min). However, this was due to that a much smaller concentration of $^3C^*$ source of DMB (15 μM) than OH source of H_2O_2 (300 μM) was used here. Direct oxidant-free photolysis was comparatively slow (~12% degradation in 420 min, with a rate constant of 0.433×10^{-3} min^{-1} and half-life of 1601 min). It is worthy to mention that aqueous oxidation of DBT could also occur under dark conditions. Two dark experiments had similar decay rates (0.532×10^{-3} min^{-1} with DMB, and 0.602×10^{-3} min^{-1} with H_2O_2), slightly quicker than the direct photolysis rate. Such dark reaction rates were close to those determined for another PAH compound phenanthrene under the same experimental conditions in our previous work [17]. Dark reactions might be caused by oxidation with dissolved O_2. In the future, we will bubble the solution before dark aging experiments to remove potential oxidation by O_2, and investigate again the dark reactions.

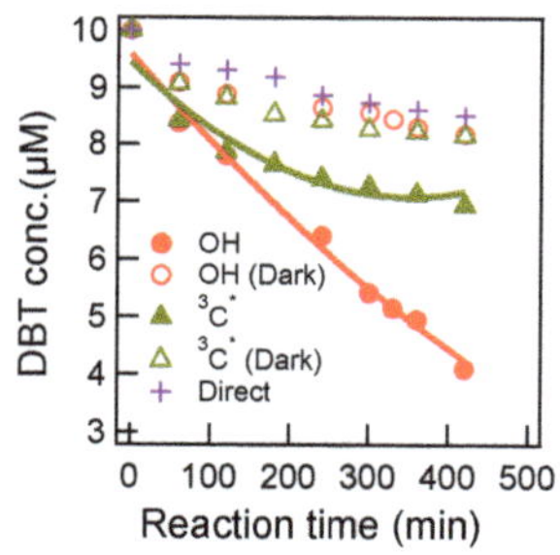

Figure 1. Concentrations of the precursor (dibenzothiophene, DBT) during aqueous oxidation under different conditions.

Table 1. Experimental conditions and the decay parameters of DBT.

Precursor	Oxidant	Oxidant Concentration	Illumination	k (10^{-3} min^{-1})	$t_{1/2}$(min)
DBT (10μM)	H_2O_2	300 μM	√	2.03 (±0.08)	342
	H_2O_2	300 μM	–	0.602 (±0.06)	1152
	DMB	15 μM	√	1.03 (±0.1)	668
	DMB	15 μM	–	0.532 (±0.08)	1303
	–	–	√	0.433 (±0.04)	1601

In Figure 2a, we present the aqSOA mass yields during photochemical oxidation of DBT by OH based on online SP-AMS measurements. The experiment lasted about 9 h, thus about two-thirds of the precursor was degraded according to the rate constant in Table 1. The mass yield was found to increase consistently during oxidation and reached ~32% till the end of illumination. It should be noted that the SP-AMS measured mass of organics at the starting point, namely $(aqOrg)_{0,\ AMS}$ in Equation (4), was significantly larger than zero. This was due to that the precursor DBT might not evaporate completely during nebulization, and some background organic impurities could remain in particle-phase as well. The initial organic mass was subtracted to calculate the aqSOA yield according to Equation (4) and it was ~20% at the precursor's half-life. As a comparison, we also determined the corresponding aqSOA yields under dark experiment and showed the results in Figure 2b. It seemed that there was no observable increase during the first 250 min (note we prepared the solution and placed it into the reaction vessel as quickly as possible to avoid dark reactions before the test); however, the aqSOA yield increased to ~15% at the end of oxidation, when ~28% DBT was degraded based on the rate expression in Table 1. The measured yields under both light irradiation and dark conditions highlight the significance of SOA formed in aqueous phase from DBT.

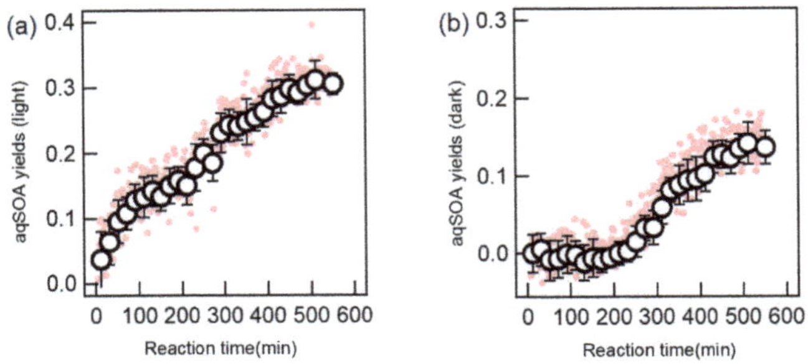

Figure 2. Mass yields of the aqSOA formed during hydroxyl radical (OH)-mediated photochemical oxidation (**a**) and dark oxidation (**b**) (note in each plot, the first pink point is zero; the first black circle is the average of the first 7 pink points, namely the 10.5 min average value; other black circles are average values of every 12 pink points, namely the 18 min average; the error bar is one standard deviation).

3.2. Chemical Properties and Evolution of aqSOA

During the OH-initiated oxidation experiment, we added ammonium sulfate as an internal standard. It is necessary to check whether or not ammonium sulfate can influence the aqueous oxidation. In this work, we found that sulfate concentration remained the same throughout the course of oxidation, and no significant N-containing fragment ions were found in the aqSOA mass spectra, indicating that both ammonium and sulfate influences were negligible. In fact, ammonium sulfate was used as an internal standard in a number of previous studies [14,24–26], and reactions between ammonium sulfate and the precursors were also not found. However, some organic precursors, such as isoprene epoxydiol [34], glycolaldehyde [37], methylglyoxal [38,39], aldehydes [40] and glyoxal [41,42], could react with ammonium. Therefore, interactions between ammonium salts and different organic species in aqueous phases and influences on the SOA are yet to be investigated.

With the SP-AMS, we could obtain an ion-speciated mass spectrum of the organic species from nebulized solution (aqOrg) every 1.5 min. Examples of such mass spectra at 60 min, 300 min and 480 min are shown in Figure 3. Ion fragments with *m/z* up to 300 amu were fitted. Based on the mass spectra, elemental ratios of the aqOrg measured at each time step were calculated by using the method of Aiken et al [43]. It should be noted that the aqOrg mass at the starting point of reaction (time *0*) was not zero; therefore, the SP-AMS determined aqOrg composition might be influenced by unevaporated precursor and background organic impurities. In this case, relative changes rather than absolute values of bulk properties of aqOrg can better reflect the evolution of aqSOA. The relative values were calculated by subtracting the absolute value at time *0* from the corresponding value at time *t*.

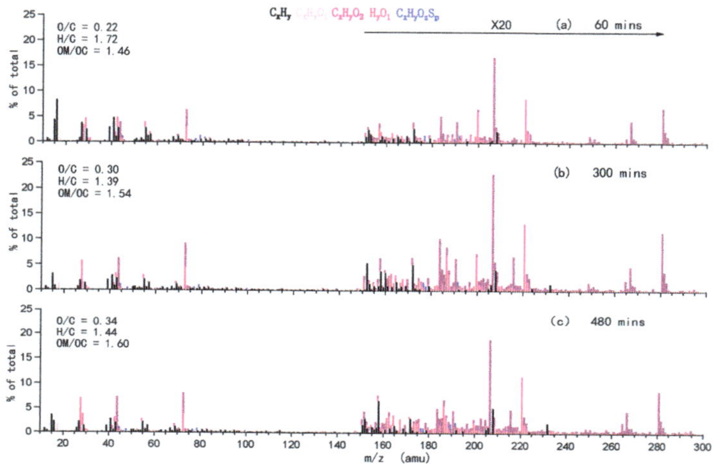

Figure 3. Examples of high-resolution mass spectra of the organics determined by the soot particle aerosol mass spectrometer (SP-AMS) during OH-mediated aqueous oxidation at different times: (**a**) 60 min, (**b**) 300min, and (**c**) 480 min.

We illustrated the relative changes of the oxygen-to-carbon (O/C) ratio (ΔO/C) and the hydrogen-to-carbon (H/C) ratio (ΔH/C) for both photochemical and dark oxidations in Figure 4. Note the absolute O/C and H/C values of the starting aqOrg were 0.14 and 1.38, respectively; while O/C and H/C of pure DBT were 0 and 0.73 (if adjusted by the AMS calibration ratio). Therefore, the starting aqOrg did contain background organics. Nevertheless, during the course of photo-oxidation, we observed a generally continuous increase of O/C and a decrease of H/C. In the end, the increment of O/C was ~0.21, while the decrease of H/C was about -0.3. This result suggests significant oxygen-addition reactions as well as a loss of species with high H/C ratios during photochemical aging. During dark oxidation, the data was largely scattered. In general, no significant changes in O/C were found. Corresponding ΔH/C ratio showed no obvious changes on average in the first 250 min but decreased to about -0.04 in the end. Note the changes of H/C matched with those of aqSOA yield, suggesting the formation of products with relatively low H/C ratios after 250 min. The oxidation state, calculated as 2 × O/C − H/C [44], is often used as a metric of the oxidation state of OA. The relative changes of OS_C (ΔOS_C, also calculated by subtraction of absolute OS_C value at time *0* from that at time *t*) were presented in Figure 5 for both photochemical and dark aged aqSOA as well. In general, ΔOS_C rose significantly to ~0.8 till the end of oxidation, while during the dark aging process it had very large uncertainties, and there was no very clear change after 250 min as those in Figures 1b and 4d. Nevertheless, degradation of DBT did occur in the first 250 min (Figure 2b), yet led to no changes in aqSOA yields, O/C, H/C and

OS_c. The underlying mechanism of the dark aging process cannot be well resolved by current SP-AMS measurements and should be investigated in the future.

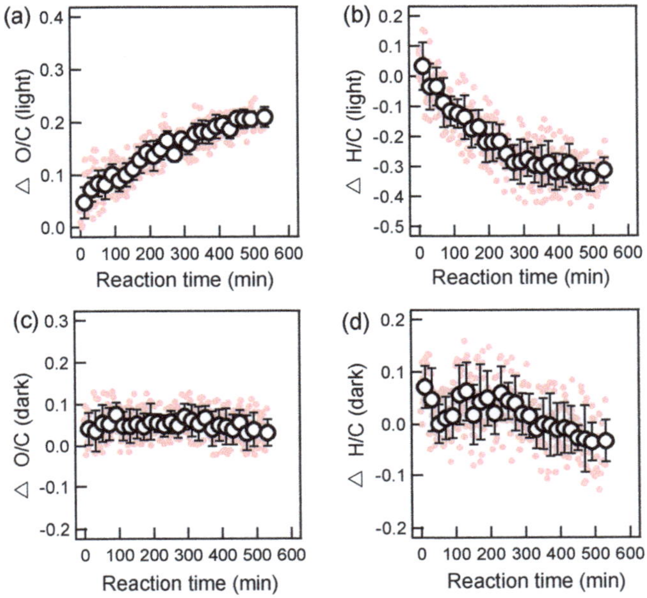

Figure 4. Time series of changes of oxygen-to-carbon (O/C) and hydrogen-to-carbon (H/C) during OH-mediated photochemical oxidation (**a**,**b**) and dark oxidation (**c**,**d**) (note in each plot, the first pink point is zero; the first black circle is the average of the first 7 pink points, namely the 10.5 min average value; other black circles are average values of every 12 pink points, namely the 18 min average; the error bar is one standard deviation).

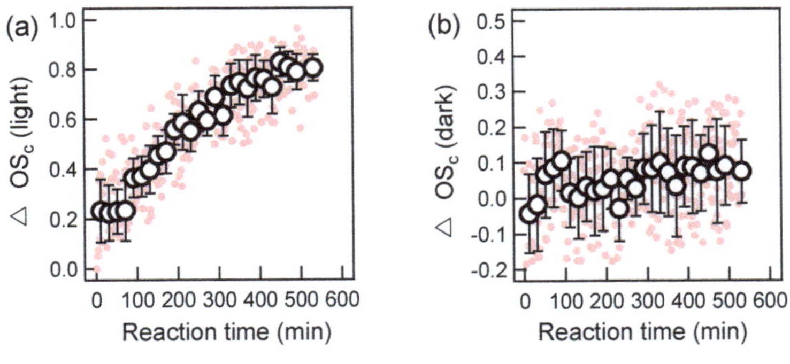

Figure 5. Temporal variations the oxidation states (OS_C) during OH-mediated photochemical oxidation (**a**) and dark oxidation (**b**) (note in each plot, the first pink point is zero; the first black circle is the average of the first 7 pink points, namely the 10.5 min average value; other black circles are average values of every 12 pink points, namely the 18 min average; the error bar is one standard deviation).

To further elaborate the aqueous oxidation mechanisms, we performed PMF analyses on the aqOrg data matrix. Similar to our and some other previous works [14,17,36], a three-factor solution was resolved. The temporal variations of different factors during the course of photochemical oxidation are shown in Figure 6, and the corresponding factor profiles are presented in Figure 7. Since the aqOrg data matrix obtained under dark had very large uncertainties as can be seen in Figure 4c,d and

Figure 5b, PMF analysis was technically impractical and the results were non-robust, therefore they are not discussed here. In fact, the results in Figure 6 had some uncertainties as well due to the very fine time resolution of the SP-AMS measurement. Factor 1, representing the first-generation products, had the smallest OS_c (−1.23). Compared with other factors, the starting concentration of Factor 1 was not zero, indicating that this factor also contained unevaporated precursor and background organics. The time series of Factor 1 increased most rapidly among three factors in the initial stage, and reached its maximum mass concentration at ~100 min, then decreased quickly and was completely depleted at ~400 min. Factor 2, with a medium OS_c (−0.97), lagged behind Factor 1, reached its peak at ~250 min, then gradually decreased to about zero till the end of oxidation (~540 min). Such behaviors suggest there was very likely a chemical transformation of Factor 1 (first-generation products) to Factor 2 (second-generation products). Factor 3 increased continuously throughout the illumination, and it has the highest OS_c (−0.70) among the three factors. Both Factor 1 and Factor 2 initially formed could contribute to Factor 3 during the oxidation. It should be noted that PMF factors can be used to describe the average particle composition during specific time periods of the experiment [45], but they may not be directly attributed to distinct chemical groups and oxidation generations. A better way to elucidate the aqSOA evolution can be fulfilled by investigating the temporal variations of tracer compounds generated from different reaction pathways. In this regard, molecular identification and quantification of the products sampled during different stages of oxidation are needed.

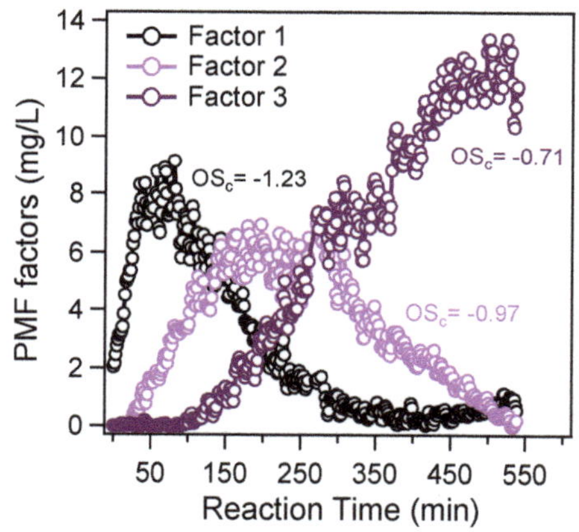

Figure 6. Temporal variations of the positive matrix factorization (PMF)-resolved factors during the course of OH-mediated photochemical oxidation.

3.3. Optical Properties of Illuminated Solutions

To probe whether or not aqueous oxidation can enhance the light absorption of DBT, we measured light absorption spectra of the starting pure DBT solution, and the final solutions upon direct photolysis, OH-oxidation and $^3C^*$-oxidation, respectively (Figure 8a). Results show that aqueous oxidation under different conditions all can increase light absorption of the precursor. Direct illuminated solution, however, mainly led to appreciable increase in the short-wavelength ultraviolet light range (<300 nm), while illuminated solution by OH also had significant enhancement over 300 nm. A very large absorption enhancement below 350 nm was observed for $^3C^*$-oxidized solution, but this was mainly due to light absorption by $^3C^*$ itself, rather than the aqSOA products; light enhancement above 350 nm due to products was in fact small.

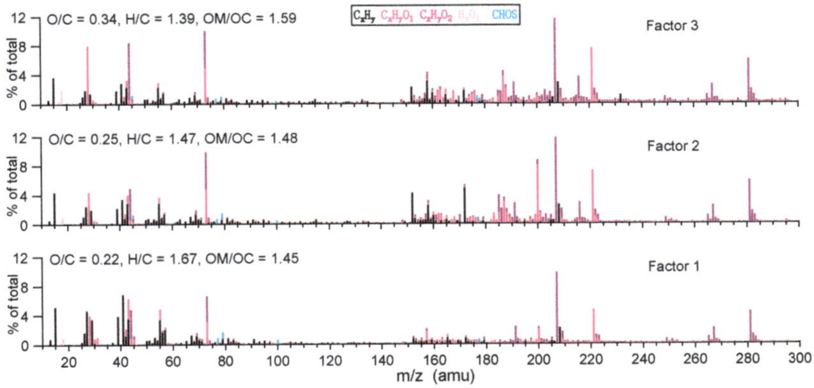

Figure 7. High-resolution mass spectra of the PMF-resolved factors (factors 1 to 3 are the names of resolved factors).

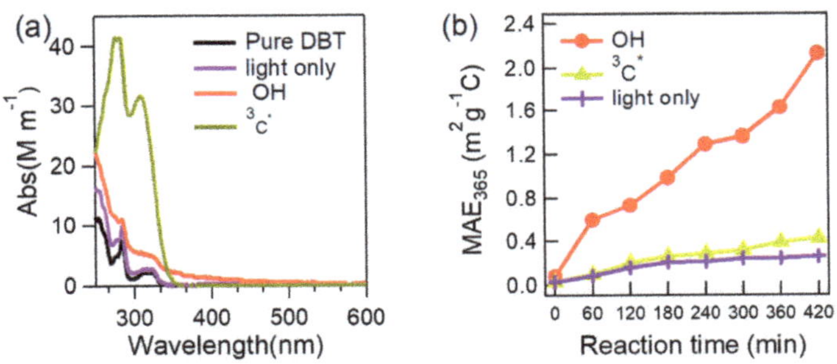

Figure 8. Light absorption spectra of the final illuminated solutions (**a**), and temporal variations of mass absorption efficiency at 365 nm (MAE_{365}) (**b**) during illumination under different conditions.

We further used the mass (using OC mass) normalized absorption coefficient at 365 nm, also known as MAE_{365}, to assess the light-absorbing ability of aqueous-phase processed solutions. Corresponding to Figure 1, MAE_{365} values of solutions during different times were determined and shown in Figure 8b. MAE_{365} was found to increase in all processes, indicating aqueous oxidation leads to BrC formation. The magnitude of the enhancement by direct light irradiation (up to 0.26 $m^2\ g^{-1}C$), and the $^3C^*$-mediated reactions (up to 0.42 $m^2\ g^{-1}C$), were relatively small. Their MAE_{365} values were lower than those observed in urban areas [29]; however, remarkable photo-enhancement was found during the course of OH-initiated photo-oxidation (up to 2.13 $m^2\ g^{-1}C$), suggesting aqueous reactions by OH can be a significant source of BrC.

4. Conclusions and Implications

This work conducted aqueous oxidation experiments on an IVOCs model compound, DBT. We showed that DBT decayed under both OH and organic triplet excited state ($^3C^*$) oxidations. Direct light irradiation could lead to photo-degradation too but to a much lesser extent. Based on our experimental condition, DBT could degrade under dark, with rate constants even higher than that of direct photolysis, yet more tests are still needed to verify the significance of dark oxidation. In the future, second-order rate constants should be investigated for a better evaluation of the decay kinetics. The aqSOA yields of OH-initiated photo-oxidation and dark oxidation were determined to be ~32% and ~15%, respectively. The oxidation degree of aqSOA increased remarkably during OH-mediated oxidation, but insignificantly during dark aging. However, the dark oxidation might be influenced by the presence of molecular O_2, and control experiments where oxygen is removed from the system should be investigated to reveal the detailed mechanism of dark oxidation. PMF analyses resolved three factors of the aqSOA from OH photo-oxidation. It was found that there was an initial formation of relatively less oxygenated first- and second-generation products, and they could chemically transform into highly oxygenated third-generation products in different stages of oxidation. In addition, light absorption measurements showed OH-mediated aqueous oxidation was a significant source of BrC.

Overall, our findings in this work demonstrate that aqueous oxidation of IVOCs (photochemical or dark aging) can be a source of SOA, pointing out a likely additional SOA source not previously considered. Our experiments mimicked the cloudy/foggy conditions. Similar processes may occur in wet aerosols but require further investigations. Of course, studies on more IVOCs with diverse saturation vapor pressures, water solubility as well as molecular structures, are highly needed to obtain a full understanding of the probabilities and significances of IVOCs-aqSOA. If it is important, aqueous processing of IVOCs should be implemented in atmospheric models to facilitate accurate simulation of SOA. In addition, our work shows that IVOCs-aqSOA (especially from OH-oxidation) is a source of BrC, therefore this process may play a role in aerosol light absorption, radiative forcing, as well as climate change.

Author Contributions: Conceptualization, Z.Y. and X.G.; data curation, Y.C. and X.G.; Formal analysis, Y.L. (Yu Liu), J.L., Y.C., Y.L. (Yue Liu) and Z.Y.; funding acquisition, X.G.; methodology, Y.L. (Yu Liu), J.L., Y.C. and Z.Y.; project administration, X.G.; writing—original draft, Y.L. (Yu Liu); writing—review and editing, X.G. All authors have read and agreed to the published version of the manuscript.

Funding: This research was funded by the Natural Science Foundation of China (NSFC) (grant number 91544220 and 21976093), and the Jiangsu Natural Science Foundation (grant number BK20181476).

Conflicts of Interest: The authors declare no conflicts of interest.

References

1. Hallquist, M.; Wenger, J.C.; Baltensperger, U.; Rudich, Y.; Simpson, D.; Claeys, M.; Dommen, J.; Donahue, N.M.; George, C.; Goldstein, A.H.; et al. The formation, properties and impact of secondary organic aerosol: Current and emerging issues. *Atmos. Chem. Phys.* **2009**, *9*, 5155–5236. [CrossRef]
2. Barsanti, K.C.; Kroll, J.H.; Thornton, J.A. Formation of Low-Volatility Organic Compounds in the Atmosphere: Recent Advancements and Insights. *J. Phys. Chem. Lett.* **2017**. [CrossRef]
3. Ervens, B.; Turpin, B.J.; Weber, R.J. Secondary organic aerosol formation in cloud droplets and aqueous particles (aqSOA): A review of laboratory, field and model studies. *Atmos. Chem. Phys.* **2011**, *11*, 11069–11102. [CrossRef]
4. Ge, X.; Zhang, Q.; Sun, Y.; Ruehl, C.R.; Setyan, A. Effect of aqueous-phase processing on aerosol chemistry and size distributions in Fresno, California, during wintertime. *Environ. Chem.* **2012**, *9*, 221–235. [CrossRef]
5. Presto, A.A.; Hennigan, C.J.; Nguyen, N.T.; Robinson, A.L. Determination of Volatility Distributions of Primary Organic Aerosol Emissions from Internal Combustion Engines Using Thermal Desorption Gas Chromatography Mass Spectrometry. *Aerosol Sci. Technol.* **2012**, *46*, 1129–1139. [CrossRef]
6. Zhao, Y.; Nguyen, N.T.; Presto, A.A.; Hennigan, C.J.; May, A.A.; Robinson, A.L. Intermediate Volatility Organic Compound Emissions from On-Road Diesel Vehicles: Chemical Composition, Emission Factors, and Estimated Secondary Organic Aerosol Production. *Environ. Sci. Technol.* **2015**, *49*, 11516–11526. [CrossRef] [PubMed]
7. Drozd, G.T.; Zhao, Y.; Saliba, G.; Frodin, B.; Maddox, C.; Chang, M.C.O.; Maldonado, H.; Sardar, S.; Weber, R.J.; Robinson, A.L.; et al. Detailed Speciation of Intermediate Volatility and Semivolatile Organic Compound Emissions from Gasoline Vehicles: Effects of Cold Starts and Implications for Secondary Organic Aerosol Formation. *Environ. Sci. Technol.* **2018**, *53*, 1706–1714. [CrossRef] [PubMed]
8. Huang, C.; Hu, Q.; Li, Y.; Tian, J.; Ma, Y.; Zhao, Y.; Feng, J.; An, J.; Qiao, L.; Wang, H.; et al. Intermediate Volatility Organic Compound Emissions from a Large Cargo Vessel Operated under Real-World Conditions. *Environ. Sci. Technol.* **2018**, *52*, 12934–12942. [CrossRef]

9. Cross, E.S.; Hunter, J.F.; Carrasquillo, A.J.; Franklin, J.P.; Herndon, S.C.; Jayne, J.T.; Worsnop, D.R.; Miake-Lye, R.C.; Kroll, J.H. Online measurements of the emissions of intermediate-volatility and semi-volatile organic compounds from aircraft. *Atmos. Chem. Phys.* **2013**, *13*, 7845–7858. [CrossRef]
10. Chan, A.W.H.; Kreisberg, N.M.; Hohaus, T.; Campuzano-Jost, P.; Zhao, Y.; Day, D.A.; Kaser, L.; Karl, T.; Hansel, A.; Teng, A.P.; et al. Speciated measurements of semivolatile and intermediate volatility organic compounds (S/IVOCs) in a pine forest during BEACHON-RoMBAS 2011. *Atmos. Chem. Phys.* **2016**, *16*, 1187–1205. [CrossRef]
11. Li, W.; Li, L.; Chen, C.-l.; Kacarab, M.; Peng, W.; Price, D.; Xu, J.; Cocker, D.R. Potential of select intermediate-volatility organic compounds and consumer products for secondary organic aerosol and ozone formation under relevant urban conditions. *Atmos. Environ.* **2018**, *178*, 109–117. [CrossRef]
12. Zhao, Y.; Hennigan, C.J.; May, A.A.; Tkacik, D.S.; de Gouw, J.A.; Gilman, J.B.; Kuster, W.C.; Borbon, A.; Robinson, A.L. Intermediate-volatility organic compounds: a large source of secondary organic aerosol. *Environ. Sci. Technol.* **2014**, *48*, 13743–13750. [CrossRef] [PubMed]
13. Zhao, Y.; Lambe, A.T.; Saleh, R.; Saliba, G.; Robinson, A.L. Secondary Organic Aerosol Production from Gasoline Vehicle Exhaust: Effects of Engine Technology, Cold Start, and Emission Certification Standard. *Environ. Sci. Technol.* **2018**, *52*, 1253–1261. [CrossRef] [PubMed]
14. Huang, D.D.; Zhang, Q.; Cheung, H.H.Y.; Yu, L.; Zhou, S.; Anastasio, C.; Smith, J.D.; Chan, C.K. Formation and Evolution of aqSOA from Aqueous-Phase Reactions of Phenolic Carbonyls: Comparison between Ammonium Sulfate and Ammonium Nitrate Solutions. *Environ. Sci. Technol.* **2018**, *52*, 9215–9224. [CrossRef]
15. Grossman, J.N.; Stern, A.P.; Kirich, M.L.; Kahan, T.F. Anthracene and pyrene photolysis kinetics in aqueous, organic, and mixed aqueous-organic phases. *Atmos. Environ.* **2016**, *128*, 158–164. [CrossRef]
16. Ye, Z.; Qu, Z.; Ma, S.; Luo, S.; Chen, Y.; Chen, H.; Chen, Y.; Zhao, Z.; Chen, M.; Ge, X. A comprehensive investigation of aqueous-phase photochemical oxidation of 4-ethylphenol. *Sci. Total Environ.* **2019**, *685*, 976–985. [CrossRef]
17. Lu, J.; Ge, X.; Liu, Y.; Chen, Y.; Xie, X.; Ou, Y.; Ye, Z.; Chen, M. Significant secondary organic aerosol production from aqueous-phase processing of two intermediate volatility organic compounds. *Atmos. Environ.* **2019**, *211*, 63–68. [CrossRef]
18. Craven, J.S.; Yee, L.D.; Ng, N.L.; Canagaratna, M.R.; Loza, C.L.; Schilling, K.A.; Yatavelli, R.L.N.; Thornton, J.A.; Ziemann, P.J.; Flagan, R.C.; et al. Analysis of secondary organic aerosol formation and aging using positive matrix factorization of high-resolution aerosol mass spectra: application to the dodecane low-NOx system. *Atmos. Chem. Phys.* **2012**, *12*, 11795–11817. [CrossRef]
19. Ho, T.C. Deep HDS of diesel fuel: chemistry and catalysis. *Catalysis Today* **2004**, *98*, 3–18. [CrossRef]
20. Hays, M.D. Nature of unresolved complex mixture in size-distributed emissions from residential wood combustion as measured by thermal desorption-gas chromatography-mass spectrometry. *J. Geophys. Res.* **2004**, *109*. [CrossRef]
21. *Appendix c. List of Determinands Assessed*; Europen Environment Agency: Copenhagen, Denmark, 2008.
22. Chao, J.; Lin, C.T.; Chung, T.H. Vapor Pressure of Coal Chemicals. *J. Phys. Chem. Ref. Data* **1983**, *12*, 1033–1063. [CrossRef]
23. Chen, H.; Ge, X.; Ye, Z. Aqueous-Phase Secondary Organic Aerosol Formation Via Reactions with Organic Triplet Excited States—a Short Review. *Curr. Pollut. Rep.* **2018**, *4*, 8–12. [CrossRef]
24. Yu, L.; Smith, J.; Laskin, A.; Anastasio, C.; Laskin, J.; Zhang, Q. Chemical characterization of SOA formed from aqueous-phase reactions of phenols with the triplet excited state of carbonyl and hydroxyl radical. *Atmos. Chem. Phys.* **2014**, *14*, 13801–13816. [CrossRef]
25. Li, Y.J.; Huang, D.D.; Cheung, H.Y.; Lee, A.K.Y.; Chan, C.K. Aqueous-phase photochemical oxidation and direct photolysis of vanillin—A model compound of methoxy-phenols from biomass burning. *Atmos. Chem. Phys.* **2014**, *14*, 2871–2885. [CrossRef]
26. Yu, L.; Smith, J.; Laskin, A.; George, K.M.; Zhang, Q. Molecular transformations of phenolic SOA during photochemical aging in the aqueous phase: Competition among oligomerization, functionalization, and fragmentation. *Atmos. Chem. Phys.* **2016**, *16*, 4511–4527. [CrossRef]

27. Collett, J.L.; Hoag, K.J.; Sherman, D.E.; Bator, A.; Richards, L.W. Spatial and temporal variations in San Joaquin Valley fog chemistry. *Atmos. Environ.* **1998**, *33*, 129–140. [CrossRef]
28. Hecobian, A.; Zhang, X.; Zheng, M.; Frank, N.; Edgerton, E.S.; Weber, R.J. Water-Soluble Organic Aerosol material and the light-absorption characteristics of aqueous extracts measured over the Southeastern United States. *Atmos. Chem. Phys.* **2010**, *10*, 5965–5977. [CrossRef]
29. Chen, Y.; Ge, X.; Chen, H.; Xie, X.; Chen, Y.; Wang, J.; Ye, Z.; Bao, M.; Zhang, Y.; Chen, M. Seasonal light absorption properties of water-soluble brown carbon in atmospheric fine particles in Nanjing, China. *Atmos. Environ.* **2018**, *187*, 230–240. [CrossRef]
30. Onasch, T.B.; Trimborn, A.; Fortner, E.C.; Jayne, J.T.; Kok, G.L.; Williams, L.R.; Davidovits, P.; Worsnop, D.R. Soot Particle Aerosol Mass Spectrometer: Development, Validation, and Initial Application. *Aerosol Sci. Technol.* **2012**, *46*, 804–817. [CrossRef]
31. Wang, J.; Ge, X.; Chen, Y.; Shen, Y.; Zhang, Q.; Sun, Y.; Xu, J.; Ge, S.; Yu, H.; Chen, M. Highly time-resolved urban aerosol characteristics during springtime in Yangtze River Delta, China: insights from soot particle aerosol mass spectrometry. *Atmos. Chem. Phys.* **2016**, *16*, 9109–9127. [CrossRef]
32. Ge, X.; Li, L.; Chen, Y.; Chen, H.; Wu, D.; Wang, J.; Xie, X.; Ge, S.; Ye, Z.; Xu, J. Aerosol characteristics and sources in Yangzhou, China resolved by offline aerosol mass spectrometry and other techniques. *Environ. Pollution* **2017**, *225*, 74–85. [CrossRef] [PubMed]
33. Ulbrich, I.M.; Canagaratna, M.R.; Zhang, Q.; Worsnop, D.R.; Jimenez, J.L. Interpretation of organic components from Positive Matrix Factorization of aerosol mass spectrometric data. *Atmos. Chem. Phys.* **2009**, *9*, 2891–2918. [CrossRef]
34. Riva, M.; Chen, Y.; Zhang, Y.; Lei, Z.; Olson, N.E.; Boyer, H.C.; Narayan, S.; Yee, L.D.; Green, H.S.; Cui, T.; et al. Increasing Isoprene Epoxydiol-to-Inorganic Sulfate Aerosol Ratio Results in Extensive Conversion of Inorganic Sulfate to Organosulfur Forms: Implications for Aerosol Physicochemical Properties. *Environ. Sci. Technol.* **2019**, *53*, 8682–8694. [CrossRef] [PubMed]
35. Zhang, Q.; Alfarra, M.R.; Worsnop, D.R.; Allan, J.D.; Coe, H.; Canagaratna, M.R.; Jimenez, J.L. Deconvolution and Quantification of Hydrocarbon-like and Oxygenated Organic Aerosols Based on Aerosol Mass Spectrometry. *Environ. Sci. Technol.* **2005**, *39*, 4938–4952. [CrossRef]
36. Zhao, R.; Aljawhary, D.; Lee, A.K.Y.; Abbatt, J.P.D. Rapid Aqueous-Phase Photooxidation of Dimers in the α-Pinene Secondary Organic Aerosol. *Environ. Sci. Technol. Lett.* **2017**, *4*, 205–210. [CrossRef]
37. Yi, Y.; Cao, Z.; Zhou, X.; Xue, L.; Wang, W. Formation of aqueous-phase secondary organic aerosols from glycolaldehyde and ammonium sulfate/amines: A kinetic and mechanistic study. *Atmos. Environ.* **2018**, *181*, 117–125. [CrossRef]
38. Aiona, P.K.; Lee, H.J.; Leslie, R.; Lin, P.; Laskin, A.; Laskin, J.; Nizkorodov, S.A. Photochemistry of Products of the Aqueous Reaction of Methylglyoxal with Ammonium Sulfate. *Acs Earth Space Chem.* **2017**, *1*, 522–532. [CrossRef]
39. De Haan, D.O.; Hawkins, L.N.; Welsh, H.G.; Pednekar, R.; Casar, J.R.; Pennington, E.A.; de Loera, A.; Jimenez, N.G.; Symons, M.A.; Zauscher, M.D. Brown carbon production in ammonium- or amine-containing aerosol particles by reactive uptake of methylglyoxal and photolytic cloud cycling. *Environ. Sci. Technol.* **2017**, *51*, 7458–7466. [CrossRef]
40. Galloway, M.M.; Powelson, M.H.; Sedehi, N.; Wood, S.E.; Millage, K.D.; Kononenko, J.A.; Rynaski, A.D.; De Haan, D.O. Secondary Organic Aerosol Formation during Evaporation of Droplets Containing Atmospheric Aldehydes, Amines, and Ammonium Sulfate. *Environ. Sci. Technol.* **2014**, *48*, 14417–14425. [CrossRef]
41. Galloway, M.M.; Chhabra, P.S.; Chan, A.W.H.; Surratt, J.D.; Flagan, R.C.; Seinfeld, J.H.; Keutsch, F.N. Glyoxal uptake on ammonium sulphate seed aerosol: reaction products and reversibility of uptake under dark and irradiated conditions. *Atmos. Chem. Phys.* **2009**, *9*, 3331–3345. [CrossRef]
42. Kampf, C.J.; Jakob, R.; Hoffmann, T. Identification and characterization of aging products in the glyoxal/ammonium sulfate system-implications for light-absorbing material in atmospheric aerosols. *Atmos. Chem. Phys.* **2012**, *12*, 6323–6333. [CrossRef]
43. Aiken, A.C.; Decarlo, P.F.; Kroll, J.H.; Worsnop, D.R.; Huffman, J.A.; Docherty, K.S.; Ulbrich, I.M.; Mohr, C.; Kimmel, J.R.; Sueper, D.; et al. O/C and OM/OC ratios of primary, secondary, and ambient organic aerosols with high-resolution time-of-flight aerosol mass spectrometry. *Environ. Sci. Technol.* **2008**, *42*, 4478–4485. [CrossRef] [PubMed]

44. Kroll, J.H.; Donahue, N.M.; Jimenez, J.L.; Kessler, S.H.; Canagaratna, M.R.; Wilson, K.R.; Altieri, K.E.; Mazzoleni, L.R.; Wozniak, A.S.; Bluhm, H. Carbon oxidation state as a metric for describing the chemistry of atmospheric organic aerosol. *Nature Chemi.* **2011**, *3*, 133–139. [CrossRef] [PubMed]
45. Koss, A.R.; Canagaratna, M.R.; Zaytsev, A.; Krechmer, J.E.; Breitenlechner, M.; Nihill, K.; Lim, C.; Rowe, J.C.; Roscioli, J.R.; Keutsch, F.N.; et al. Dimensionality-reduction techniques for complex mass spectrometric datasets: Application to laboratory atmospheric organic oxidation experiments. *Atmos. Chem. Phys. Discussion* **2019**, *2019*, 1–42. [CrossRef]

atmosphere

MDPI

Article

Aqueous-Phase Brown Carbon Formation from Aromatic Precursors under Sunlight Conditions

Kristijan Vidović *, Ana Kroflič, Martin Šala and Irena Grgić *

Department of Analytical Chemistry, National Institute of Chemistry, Hajdrihova 19, SI-1000 Ljubljana, Slovenia; ana.kroflic@ki.si (A.K.); martin.sala@ki.si (M.Š.)

* Correspondence: kristijan.vidovic@ki.si (K.V.); irena.grgic@ki.si (I.G.)

Received: 22 November 2019; Accepted: 22 January 2020; Published: 24 January 2020

Abstract: At present, there are still numerous unresolved questions concerning the mechanisms of light-absorbing organic aerosol (brown carbon, BrC) formation in the atmosphere. Moreover, there is growing evidence that chemical processes in the atmospheric aqueous phase can be important. In this work, we investigate the aqueous-phase formation of BrC from 3-methylcatechol (3MC) under simulated sunlight conditions. The influence of different HNO_2/NO_2^- concentrations on the kinetics of 3MC degradation and BrC formation was investigated. Under illumination, the degradation of 3MC is faster ($k_{2nd(global)} = 0.075\ M^{-1}{\cdot}s^{-1}$) in comparison to its degradation in the dark under the same solution conditions ($k_{2nd} = 0.032\ M^{-1}{\cdot}s^{-1}$). On the other hand, the yield of the main two products of the dark reaction (3-methyl-5-nitrocatechol, 3M5NC, and 3-methyl-4-nitrocatechol, 3M4NC) is low, suggesting different degradation pathways of 3MC in the sunlight. Besides the known primary reaction products with distinct absorption at 350 nm, second-generation products responsible for the absorption above 400 nm (e.g., hydroxy-3-methyl-5-nitrocatechol, 3M5NC-OH, and the oxidative cleavage products of 3M4NC) were also confirmed in the reaction mixture. The characteristic mass absorption coefficient (MAC) values were found to increase with the increase of NO_2^-/3MC concentration ratio (at the concentration ratio of 50, MAC is greater than 4 $m^2{\cdot}g^{-1}$ at 350 nm) and decrease with the increasing wavelength, which is characteristic for BrC. Yet, in the dark, roughly 50% more BrC is produced at comparable solution conditions (in terms of MAC values). Our findings reveal that the aqueous-phase processing of 3MC in the presence of HNO_2/NO_2^-, both under the sunlight and in the dark, may significantly contribute to secondary organic aerosol (SOA) light absorption.

Keywords: brown carbon chromophores; biomass burning; photooxidation; nitration; nitrophenols; methylnitrocatechols; mass absorption coefficient

1. Introduction

The existence of light-absorbing organic aerosols in the atmosphere, also termed as brown carbon (BrC), has become increasingly evident in recent years [1–3]. In contrast to black carbon (BC), which strongly absorbs light in the entire spectral range [4], BrC typically absorbs in the near-UV (300–400 nm) and blue-to-green visible ranges (400–550 nm) [1,2,5]. BrC impacts radiative forcing directly by absorbing solar irradiation and also indirectly by influencing the cloud formation [6,7]. In addition, its components can take part in different (photo)chemical reactions in the atmosphere and can also have harmful effects on the environment and living organisms [1,8].

BrC is largely observed in areas influenced by biomass and biofuel burning and is often attributed to primary emissions [1,9]. However, there is more and more evidence that secondary transformations of primarily emitted volatile organic compounds (VOCs), in the gas and/or in the atmospheric aqueous phase, also contribute to BrC formation [10–12]. Important precursors to BrC are aromatic compounds, which contribute significantly to the budget of atmospheric pollution [13]; among those, substituted aromatics are especially important [14]. A class of aromatics with strong absorption in the near UV and visible are nitrophenols (NP), which are frequently determined in atmospheric particulate matter (PM), with elevated concentrations in winter PM and being primarily associated with biomass burning (BB) [15–19]. It has recently been shown that several nitroaromatic pollutants, such as NP, nitrocatechols (NC), and methylnitrocatechols (MNC), can be formed via (photo)chemical oxidation of preferentially gas-phase precursors in the atmosphere [20,21]. Moreover, evidence exists that their formation can take place in the atmospheric aqueous phase, which additionally contributes to the atmospheric abundance of these compounds [11,22,23].

At present, there are numerous open questions concerning the mechanisms of BrC formation, particularly those in cloud droplets and aqueous particles. However, there is rising evidence that chemical processes in the atmospheric condensed phase can efficiently contribute to organic aerosol aging, and thus eventually also to light-absorbing secondary organic aerosol (SOA) formation [24,25]. Yellow-colored MNC have been recognized as essential constituents of BB SOA and further also proposed as suitable tracer compounds [15,16,22,26]. Their complex formation mechanisms under nighttime conditions in the atmospheric aqueous phase have recently been extensively studied [27,28]. Moreover, it has been demonstrated that under sunlight conditions, the formation of nitroguaiacols (nitrated 2-methoxyphenols, NG) and their aging in the atmospheric aqueous phase are even more complex [11,23]. In general, nitrophenols have been shown to be susceptible to direct photolysis and to photo-oxidation [29–31]; as a result, the light-absorbing properties of BrC can change significantly during atmospheric aging [32,33]. Hems and Abbatt [33] have recently investigated the mechanism of color enhancement and fate of nitrophenols (NC, NG, and dinitrophenol) during aqueous-phase photo-oxidation. They have found out that the initially fast-formed nitrophenols, being functionalized with additional OH groups, likely lead to the increased absorption in the visible range, whereas further reactions lead to product formation with poor or no visible absorption (i.e., in the process called bleaching).

In this work, aqueous-phase transformations of 3-methylcatechol (3MC) as an important precursor to light-absorbing MNC were investigated under atmospheric sunlight conditions. The main goal was to find out how photochemical processing affects the kinetics of 3MC degradation at different HNO_2/NO_2^- concentrations, and how the absorptive properties of the reaction mixture change during the aging. Although NO_3^- is the prevailing nitrogen species in atmospheric aerosols, HNO_2/NO_2^- is known for its much higher reactivity at ambient conditions, at least in the dark, and is thus considered important for SOA formation and aging. Special attention was paid to the main dark BrC products, 3-methyl-5-nitrocatechol (3M5NC), and 3-methyl-4-nitrocatechol (3M4NC), and to the possible formation of absorbing second-generation products, such as hydroxy-3-methyl-5-nitrocatechol (3M5NC-OH). In addition, the influence of different reaction conditions on BrC formation/decomposition was quantified by a cumulative parameter, mass absorption coefficient (MAC in $m^2\ g^{-1}$), which was used to describe the evolution of BrC during the solution aging.

2. Materials and Methods

2.1. Materials

3-methylcatechol (3MC) (purity ≥98%) and sodium nitrite ($NaNO_2$, ACS reagent, ≥97.0%), both from Sigma-Aldrich were used for the reaction mixture preparation in high purity water (18.2 MΩ cm), supplied by a Milli-Q water purification system. Standards of two reaction products, 3M5NC and 3M4NC, were prepared in the laboratory by the procedures adopted from Palumbo [34] and Kitanovski [35]. Sulfuric acid 98% (H_2SO_4, EMSURE, p.a. grade) was used for the pH adjustment.

For the quantification of 3MC, $NaNO_2$ and first-generation reaction products (3M5NC and 3M4NC), acetonitrile (Sigma-Aldrich, St. Louis, MO, USA, gradient grade, for HPLC >99.9%) and formic acid (Kemika, Zagreb, Croatia) were used for mobile phase preparation. For the second-generation product identification, methanol (Fluka, Chromosolv LC-MS grade, ≥99.9%) and an ion-pair reagent for LC-MS Dibutylammonium Acetate-DBAA (Tokyo Chemical Industry, Tokyo, Japan, 0.5 mol·L^{-1}) were used.

2.2. Laboratory Experiments

The photolysis of 3MC in mildly acidic (pH = 4–5, adjusted with H_2SO_4) aqueous solutions of sodium nitrite ($NaNO_2$;) was investigated at ambient temperature (25 °C). The reactions were carried out in a custom-built reactor, which is a modified rotating evaporator (Büchi, Flawil, Switzerland) equipped with a thermostated bath [36]. Only briefly, a low-volume condenser was installed to prevent significant evaporation from the solution ensuring a closed physical system. The reaction mixture (100 mL) was continuously mixed by rotation (50 rpm) of a round-bottom flask made of the DURAN glass (transparent one for daytime experiments and amber glass for dark reactions). For illumination, a LOT-QuantumDesign Europe solar simulator (L.O.T.-Oriel GmbH & Co. KG, Darmstadt, Germany) equipped with a xenon short-arc lamp (300 W, ozone free) was used. According to the specifications, the simulator produces irradiation equivalent to approximately one sun at the working distance of 180 mm. Moreover, the cut-off of the DURAN glass at 300 nm resembles the absorption of the stratospheric ozone, which allows us to mimic ambient conditions very well. The concentrations of 3MC and H_2SO_4 were fixed at 0.1 and 0.05 mM, respectively. The influence of $NaNO_2$ concentration was investigated at 0.1, 0.5, 1.0, 2.0, and 5.0 mM. Samples (1.5 mL aliquots) were taken from the reaction mixture according to the predefined time protocol (usually each hour). In addition, a control experiment of the direct photolysis of 3MC without the addition of $NaNO_2/H_2SO_4$, and the dark nitration of 3MC in acidic $NaNO_2$ (1 and 2 mM) solutions were performed.

An ultra-high pressure liquid chromatography (UltiMate 3000 UHPLC System; Thermo Fischer Scientific, Waltham, MA, USA) coupled with a diode array detector (DAD) or a triple quadrupole/linear ion trap mass spectrometer (4000 Q TRAP LC-MS/MS System; Applied Biosystems/MDS Sciex) was used for the detection of reactants and products. For the quantification of 3MC, $NaNO_2$, and first-generation products (3M5NC and 3M4NC), an isocratic elution program with acetonitrile/0.1% formic acid (70/30, V/V) at the flow rate of 0.6 mL·min^{-1} was applied, assuring the separation of components on an Atlantis T3 column (3.0 × 150 mm^2, 3 μm particle size; Waters, Milford, MA, USA). The injection volume and the column temperature were 10 μL and 30 °C. The detection wavelengths used in the case of HPLC-DAD measurements were 275 nm for 3MC, 345 nm for MNC, and 355 nm for $NaNO_2$. For the second-generation product identification (*m/z* 182 and 184), different chromatographic conditions were used: a Hypersil GOLD aQ column (2.1 × 150 mm^2, 3 μm particle size; Thermo Scientific, Waltham, MA, USA) and an isocratic elution with methanol/50 mM DBAA (10/90, V/V) at the flow rate of 0.3 mL·min^{-1}. The injection volume and the column temperature were again 10 μL and 30 °C. The detection wavelength used in this case was 388 nm.

Absorption spectra of the reaction solution were measured offline in a 1 cm quartz cuvette with a UV-Vis spectrometer (Lambda 25, PerkinElmer, Waltham, MA, USA) immediately after sampling. The absorbance was monitored in a broad spectral range from 200 to 700 nm.

2.3. Kinetic Analysis

The experimentally obtained 3MC photolysis profiles were first treated by a pseudo-first-order kinetics:

$$\frac{\partial c}{\partial t} = -k_{app} \cdot c, \quad (1a)$$

where c is the concentration of 3MC, t denotes time, and k_{app} is the apparent pseudo-first-order kinetic rate constant (in s^{-1}) of 3MC degradation. The integrated form of Equation (1a) was fitted to the experimental data points, c_0 being the initial concentration of 3MC in the reaction mixture.

$$c = c_0 \cdot e^{-k_{app} \cdot t}. \quad (1b)$$

Referring to the direct photolysis of 3MC, which is negligible (see the control experiment), we postulated that the degradation of 3MC be dependent on the concentration of $NaNO_2$.

$$\frac{\partial c}{\partial t} = -k_{2nd} \cdot c(NaNO_2) \cdot c. \tag{2a}$$

Except for the condition where the initial concentrations of 3MC and $NaNO_2$ were equal (i.e., 0.1 mM; experiment 1), we further assumed constant $NaNO_2$ concentration in the reaction mixture throughout the course of experiment. Second-order kinetic rate constants (k_{2nd} in $M^{-1} \cdot s^{-1}$) were thus estimated from k_{app} as:

$$k_{app} = k_{2nd} \cdot c(NaNO_2)_0. \tag{2b}$$

$c(NaNO_2)_0$ being the initial concentration of $NaNO_2$ in the reaction mixture. Note, however, that one cannot assume constant $NaNO_2$ concentration during experiment 1. Therefore, a second-order kinetic treatment was applied in this case, taking into account the measured $c(NaNO_2)$ at each time of the experiment.

$$\frac{\partial c}{\partial t} = -k_{2nd} \cdot c(NaNO_2)_t \cdot c, \tag{3}$$

where $c(NaNO_2)_t$ represents the concentration of $NaNO_2$ at the reaction time t.

The experimental data for which the assumption of a pseudo-first-order is valid ($c(NaNO_2) = 0.5$, 1.0, 2.0, and 5.0 mM) were further treated simultaneously by the following function:

$$\frac{\partial c}{\partial t} = -k_{2nd}(\text{global}) \cdot c(NaNO_2)_0 \cdot c. \tag{4}$$

This gave us a global second-order kinetic rate constant, k_{2nd}(global) representing all treated experimental conditions. Note that parameters obtained by accounting for the data at different experimental conditions at once tend to be more universally applied.

OriginPro 2018 was used for the fitting of the model functions to the experimental data (i.e., measured concentration and absorption profiles).

2.4. Determination of Absorption Properties

The obtained k_{2nd}(global) (Equation (4)) was further used to quantitatively describe the influence of the studied reaction conditions on BrC formation. In the case of experiment 1 (Table 1), k_{2nd}(global) is not valid; therefore, k_{app} (Equation (1a)) was used instead. According to the Beer–Lambert Law, the measured absorbance of a sample at a distinct wavelength, A_{tot} is dependent on the characteristic mass absorption coefficient of contained species i at this wavelength, MAC_i, and their concentrations, c_i; l is the absorption path length characteristic of the used cuvette.

$$A_{tot} = \sum_i MAC_i \cdot c_i \cdot l. \tag{5a}$$

Table 1. Ratios and yields of quantified 3-methyl-5-nitrocatchol (3M5NC) and 3-methyl-4-nitrocatechol (3M4NC) at different $NaNO_2$ concentrations under sunlight conditions after 8 h of reaction.

Experiment	3MC [mM]	$NaNO_2$ [mM]	3M5NC/3M4NC	pH	MNC Product Yield after 8 h
1	0.1	0.1	5	3.9	0.17
2	0.1	0.5	6	4.5	0.45
3	0.1	1.0	96	4.6	0.40
4	0.1	2.0	98	4.9	0.36
5	0.1	5.0	63	5	0.27

By definition, BrC absorbs in the near-UV and visible ranges, so we attributed A_{tot} measured above 300 nm to the BrC formation (note that 3MC does not absorb in this range). Equation (5a) can be thus expressed as:

$$A_{tot}(>300\ \text{nm}) = \text{MAC}_{\text{BrC}} \cdot m_{\text{BrC}} \cdot l. \tag{5b}$$

MAC_{BrC} (in $m^2 \cdot g^{-1}$) being the mass absorption coefficient characteristic of the formed BrC and m_{BrC} is its mass concentration in g m^{-3}.

By combining Equation (5b) with the expression for the pseudo-first-order product formation and taking into account the law of conservation of mass (i.e., assuming that all consumed 3MC was converted to BrC; m_0 is the initial 3MC mass concentration):

$$m_{\text{BrC}} = m_0\left(1 - e^{-k_{2\text{nd}}(\text{global}) \cdot c(\text{NaNO}_2)_0 \cdot t}\right). \tag{6}$$

One can describe the evolution of A_{tot} with time with the following function:

$$A_{tot} = \text{MAC}_{\text{BrC}} \cdot m_0\left(1 - e^{-k_{2\text{nd}}(\text{global}) \cdot c(\text{NaNO}_2)_0 \cdot t}\right) \cdot l. \tag{7}$$

The derived function was fitted to the experimental data points, and characteristic MAC_{BrC} values for the investigated experimental conditions were obtained. Note that in the case of experiment 1 Table 1), k_{app} was used instead of the product of $k_{2nd(global)}$ and $c(NaNO_2)_0$.

3. Results and Discussion

3.1. Kinetic Analysis of 3MC Photolysis in Mildly Acidic $NaNO_2$ Solution

In the dark reaction of 3MC and $NaNO_2$, two main reaction products have been recently quantified, i.e., 3M5NC and 3M4NC; they accounted for 70–100% of the reacted 3MC at similar reaction conditions as used in this study [27]. In addition, the ratio of 3M5NC to 3M4NC was constant during the dark reaction and dependent only on the concentration of HNO_2 [27].

Under sunlight conditions, dark reaction mechanisms also exist; thus, 3M5NC and 3M4NC were expected to be formed in this study as well. The performed experiments under illumination are listed in Table 1 together with the corresponding 3M5NC/3M4NCratios and product yields determined after 8 h of reaction. The product yield was calculated as the sum of concentrations of target reaction products divided by the concentration of the reacted 3MC.

$$\text{Product yield} = \frac{\Sigma[MNC]_{8h}}{[3MC]_{0h} - [3MC]_{8h}}. \tag{8}$$

Under the sunlight, the cumulative yield of identified MNC products was only up to about 45%. The ratio (and the yield) of 3M5NC to 3M4NC were additionally found to be dependent on the initial concentration of $NaNO_2$ (Table 1). The low product yield and the variable product ratio suggest the existence of different oxidation pathways of 3MC, and formation and/or degradation pathways of MNC products in comparison to the dark reaction.

During the photolysis of HNO_2, diverse reactive species are formed (e.g., OH, NO, NO_2), which are possible sinks of 3MC [37]. In general, OH radicals react with aromatics with high (near-diffusion controlled) rate constants in the order of 10^{10} $L \cdot mol^{-1} \cdot s^{-1}$ at 298 K [38,39]. The reactivity of aromatics with NO_2 is still relatively high (in the order of 10^8–10^9 L $mol^{-1} \cdot s^{-1}$), whereas NO seems to be too weak to react with aromatic compounds directly [23]. In summary, other degradation pathways of MNC, i.e., initiated by radicals, are very likely under illumination [33].

Under illumination, also the decay of 3MC was greatly dependent on the applied conditions (Table 2). In general, k_{app} increased as the $NaNO_2$ concentration increased. Note, however, that in the cases where $NaNO_2$ was in excess, k_{2nd} slightly decreased with $NaNO_2$ addition. This may be due to unaccounted 3MC degradation pathways, independent of (or indirectly dependent on) the $NaNO_2$ concentration.

Table 2. Kinetic rate constants (apparent pseudo-first-order, k_{app}; second-order, k_{2nd}; and global second-order, k_{2nd}(global)) of the photolysis of 3-methylcatechol (3MC) in aqueous solution (pH = 4.5–5, T = 25 °C) at different initial concentrations of $NaNO_2$ under simulated sunlight conditions (YES) and in the dark (NO). The initial concentration of 3MC was always 0.1 mM.

Experiment	$NaNO_2$ [mM]	Sunlight	k_{app} (s^{-1})	k_{2nd} ($M^{-1}s^{-1}$)	k_{2nd}(global) ($M^{-1} \cdot s^{-1}$)
1	0.1	YES	7.88×10^{-6}	0.168	–
2	0.5	YES	4.47×10^{-5}	0.089	0.075
3	1.0	YES	7.30×10^{-5}	0.073	
4	2.0	YES	1.30×10^{-4}	0.065	
5	5.0	YES	2.40×10^{-4}	0.048	
6	1.0	NO	4.26×10^{-5}	0.043	0.032
7	2.0	NO	5.03×10^{-5}	0.025	

3.1.1. Kinetic Analysis at Low $NaNO_2$ Concentrations

The time-dependent concentrations of 3MC (0.1 mM initial concentration) in 0.1 mM $NaNO_2$ solution under sunlight conditions are shown in Figure 1. In the investigated time range, k_{app} (7.88×10^{-6} s^{-1}) describes the behavior of this reaction reasonably well and is comparable with the reaction in the dark (8.33×10^{-6} s^{-1}) [27]. However, the yield of identified MNC under sunlight is barely 17% after 8 h (Table 1), pointing to different degradation and/or formation pathways of 3MC and MNC, respectively.

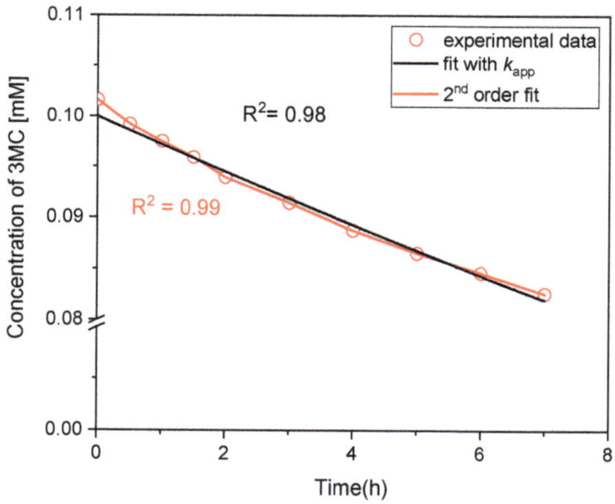

Figure 1. 3-methylcatechol (3MC) degradation in aqueous solution at 0.1 mM initial concentrations of 3MC and $NaNO_2$ (pH = 3.9, T = 25 °C). Experimental data (symbols) and calculated time-dependent concentration profiles (lines) for the pseudo-first order (k_{app}) and second order (k_{2nd}) kinetics.

Second-order kinetics was also applied to this set of experimental data, and the comparison of both models (Equations (1a) and (3)) is shown in Figure 1. As k_{app} describes experimental data adequately in the investigated time range, we used it in further calculations with a fair amount of confidence.

3.1.2. Kinetic Analysis at High $NaNO_2$ Concentrations

Time-dependent concentrations of 3MC in the presence of different initial $NaNO_2$ concentrations (i.e., from 0.5 to 5 mM) are shown in Figure 2a–d. The lines are the calculated concentration profiles according to the two proposed model functions (Equations (1a) and (4)). By using k_{app}, the experimental data are represented well in all cases; thus, its values were further used to derive k_{2nd} according to Equation (2b) for each experiment individually (see Table 2). Yet, a global fit was also performed to get a more reliable second-order rate constant, $k_{2nd(global)}$, by which the kinetics of 3MC decay at all treated conditions is still well described (black lines in Figure 2).

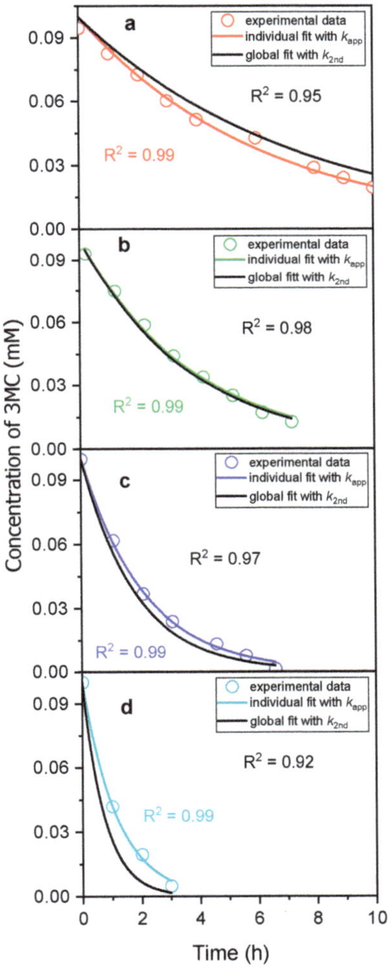

Figure 2. 3-methylcatechol (3MC) degradation in aqueous solution at different initial concentrations of $NaNO_2$ (pH = 4.5–5, T = 25 °C). The concentrations of $NaNO_2$ were 0.5 (**a**), 1.0 (**b**), 2.0 (**c**), and 5.0 mM (**d**). Experimental data (symbols) and calculated time-dependent concentration profiles (lines) for the pseudo-first order (k_{app}) and global second order (k_{2nd} (global)) fits.

The rate constant for 3MC decay in the simulated sunlight ($k_{2nd(global)} = 0.075\ M^{-1}{\cdot}s^{-1}$) is twice as high as in the dark ($k_{2nd} = 0.032\ M^{-1}{\cdot}s^{-1}$) under similar initial conditions (Table 2). As mentioned above, several different degradation mechanisms are possible in the sunlight that are prevented in the dark. Similarly, under the sunlight, k_{2nd} for the 3MC decay at the lowest $NaNO_2$ concentration (0.1 mM) is almost twice as that at higher $NaNO_2$ concentrations (0.5–5 mM). The reason might be in the competition between NO_2^- and 3MC for OH radicals and other reactive species at these experimental conditions. When HNO_2/NO_2^- is in excess, OH radicals preferentially react with NO_2^-, whereas only a minor fraction reacts with 3MC, which results in its slower degradation [37,39].

3.2. *Absorption Spectra and BrC Formation*

3.2.1. Low $NaNO_2$ Concentrations

As has recently been reported in our previous study on the nighttime aqueous-phase formation of MNC, the two identified products, i.e., 3M5NC and 3M4NC, strongly absorb in the near UV range, whereas the reaction mixture also exhibited notable absorption in the visible range, which are all characteristic of BrC [27]. Moreover, in the dark, the enhanced absorption in the visible range can be attributed to MNC degradation by hydroxylation and oxidative cleavage by water addition [28].

Figure 3a–f shows the time evolution of absorption spectra during the aging of 3MC in mildly acidic $NaNO_2$ solutions under illumination. The dependence of absorbance on the initial concentration ratio of 3MC to $NaNO_2$ was investigated. Compared to the photolysis of 3MC only (without the addition of $NaNO_2$), where we did not observe any change in the absorption spectrum over the reaction time scale (Figure 3a), the spectrum noticeably evolves already at low $NaNO_2$ concentration (i.e., 0.1 mM, Figure 3b). The absorption around 350 nm increases with time and corresponds to the formation of first-generation products, 3M5NC and 3M4NC [27]. Nevertheless, only a slight increase in the absorption is in agreement with the ascertainment that at low NO_x, BrC formation by nitration is not preferred [40,41]. Note also the low product yield at the lowest NO_2^- concentration (Table 1).

The time evolution of absorbance at 350, 400, and 410 nm, measured experimentally, and modeled using Equation (7), is presented in Figure 4. The predicted absorption describes the experimental data fairly well; a somewhat weaker correlation can be explained by extensive scattering of data due to unreliable measurements at low absorbance (below 0.2). The developed absorption above 400 nm at long reaction times may, among others, correspond to hydroxylated and ring cleavage products formed from MNC, the same as in the dark [28]. See further discussion for justification.

3.2.2. High $NaNO_2$ Concentrations

At higher $NaNO_2$/3MC concentration ratios, the absorption at 350 nm is substantially enhanced in comparison to the experiment with equimolar amounts of both reactants, resulting in the appearance of a distinctive peak with the maximum at this wavelength (Figure 3c–e). In addition, the absorption above 400 nm becomes more pronounced. The enhancement in absorption at examined wavelengths is the fastest at the highest $NaNO_2$ concentration (Figure 3f).

As already mentioned before, the two products, 3M5NC and 3M4NC with a pronounced absorption peak at 350 nm can react further, forming second-generation products, which can be responsible for the development of absorption in the visible region. According to our recent study [28], the secondary products can be 3M5NC-OH and the oxidative cleavage products of 3M4NC. Their spectroscopic characteristics (both exhibit the absorption maxima at cca. 400 nm and the latter another one at around 500 nm [28]) are in agreement with the presented results and were also confirmed in one of the investigated reaction mixtures by comparison with the recently published data. The chromatogram of the reaction mixture is shown in Figure 5 and mass spectra corresponding to the characteristic peaks were additionally analyzed by an LC-MS/MS experiment. The peak at 2.5 min showed the *m/z* ratio of 183 corresponding to 3M5NC-OH, while the peak at 4.5 min with the *m/z* of 184 corresponds to the oxidative cleavage products of 3M4NC, which additionally confirmed the identity of investigated compounds.

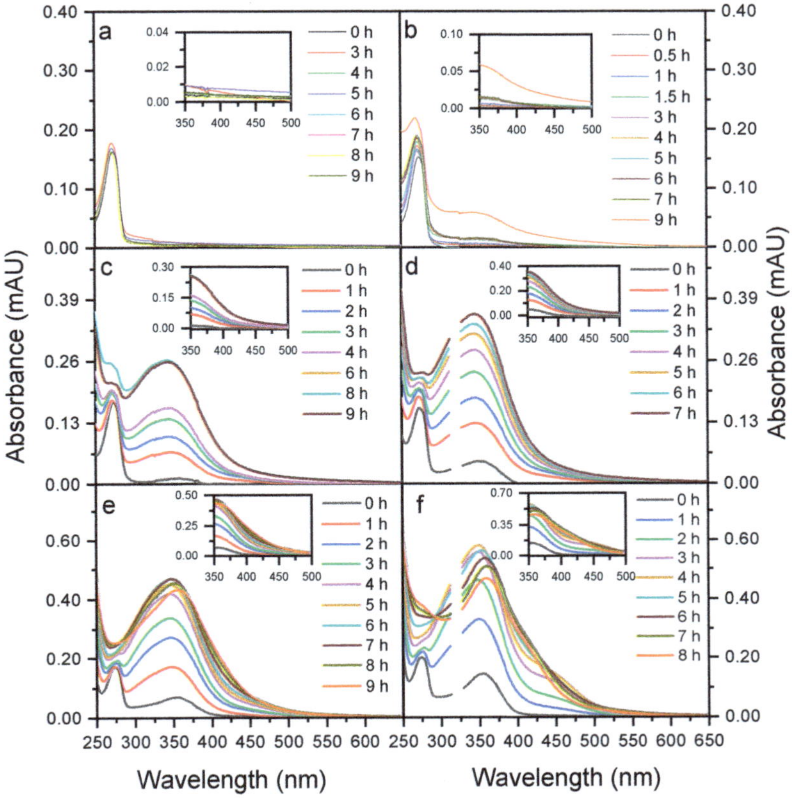

Figure 3. Absorption spectra of the reaction mixture during the photolysis of 3-methylcatechol (3MC) in aqueous $NaNO_2$ solutions (pH = 4–5, T = 25 °C) at different initial concentrations of $NaNO_2$. The concentration of 3MC was always 0.1 mM, while the concentrations of $NaNO_2$ were 0 (**a**), 0.1 (**b**), 0.5 (**c**), 1 (**d**), 2 (**e**), and 5 mM (**f**). The discontinuation in the absorption spectra (d,f) is due to the issues with the instrument and does not affect the results in any way.

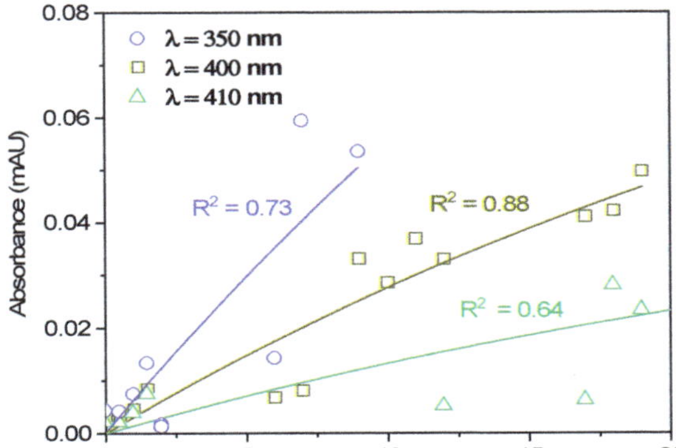

Figure 4. Brown carbon (BrC) formation in 0.1 mM 3-methylcatechol and 0.1 mM $NaNO_2$ under sunlight conditions. Experimentally measured (symbols) and modeled absorbance (solid lines) at different wavelengths representative of BrC.

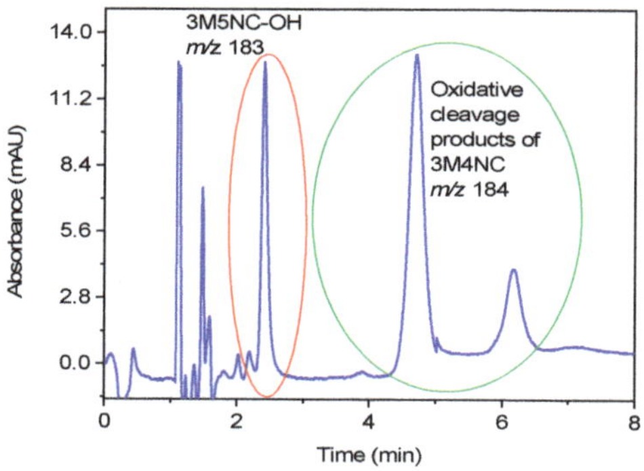

Figure 5. Chromatogram recorded at 388 nm for the reaction mixture of 0.1 mM 3MC and 1 mM $NaNO_2$ under sunlight conditions is comparable to the one from our previous study [28]. The peak at 2.5 min with the *m*/*z* ratio of 183 corresponds to 3MC5NC-OH, while the peak at 4.5 min with the *m*/*z* of 184 corresponds to oxidative cleavage products of 3M4NC.

The time evolution of absorbance, experimentally measured and modeled by using Equation (7), is presented in Figure 6 for all remaining four experimental conditions. At the initial concentrations of 0.5 and 1 mM $NaNO_2$, the experimental data are very well fitted with the applied absorption model (Figure 6a,b). However, at higher $NaNO_2$ concentrations, the exponential shape of absorbance changes into the sigmoidal one, which is most noticeable at 5 mM $NaNO_2$ in the visible range (Figure 6c,d). This behavior could be attributed to the formation of the second-generation products, which are particularly responsible for the enhanced absorption above 400 nm (e.g., 3M5NC-OH and the oxidative cleavage products of 3M4NC). As the proposed absorption model only considers the one-step transformation of 3MC into BrC, it fails to precisely reproduce such complex behavior of the experimental system. Furthermore, at the highest NO_2^- concentration, 3MC is completely consumed during the experiment; therefore, primary reaction products, 3M5NC and 3M4NC, stopped forming already before the end of the experiment. Consequently, the decay of absorbance is observed in the second part of the experiment, and the predicting ability of the absorption model is even worse in this case because it does not account for any of the explained phenomena.

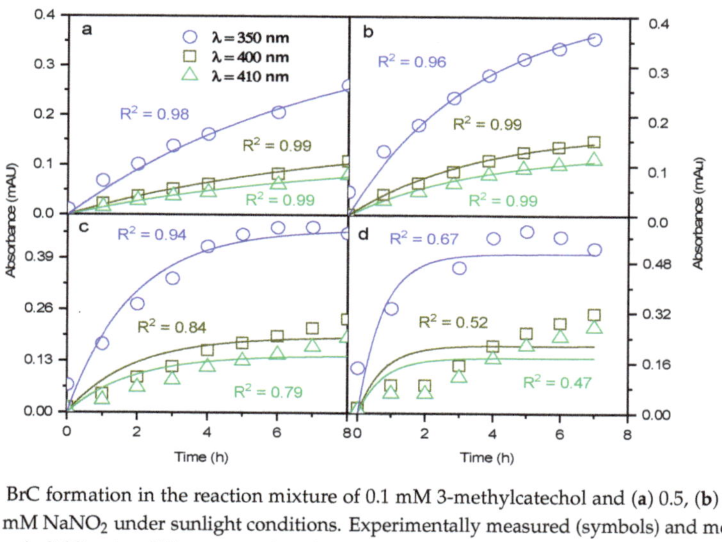

Figure 6. BrC formation in the reaction mixture of 0.1 mM 3-methylcatechol and (**a**) 0.5, (**b**) 1, (**c**) 2, and (**d**) 5 mM $NaNO_2$ under sunlight conditions. Experimentally measured (symbols) and modelled absorbance (solid lines) at different wavelengths, representative of BrC are shown.

In contrast to the illuminated experiments, the model describes the experimental data of dark reactions very well, also at high $NaNO_2$ concentrations (0.1 mM 3MC and 2 mM $NaNO_2$; Figure 7). These results additionally support our discussion that under the sunlight, the formed BrC species undergo secondary reactions substantially.

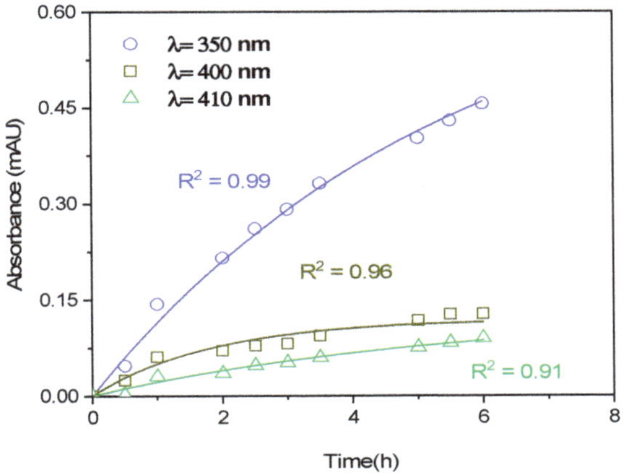

Figure 7. BrC formation in the reaction mixture of 0.1 mM 3-methylcatechol and 2 mM $NaNO_2$ in the dark. Experimentally measured (symbols) and modelled absorbance (solid lines) at different wavelengths, representative of BrC are shown.

From characteristic MAC values at 350, 400, and 410 nm listed in Table 3 one can see that with increasing $NaNO_2$ concentration, higher MAC values are obtained corresponding to more BrC production; i.e., the highest was in the order of 4 $m^2\ g^{-1}$ for 5 mM $NaNO_2$ at 350 nm. It should be emphasized that in the dark (experiments 6 and 7), roughly 50% more BrC is produced at comparable solution conditions. This is consistent with the finding that under the sunlight, BrC degradation, and so the solution bleaching, may proceed via the reactions with OH radicals leading to products with poor absorption in the visible region [33].

Table 3. Wavelength-dependent mass absorption coefficients (MAC_{350}, MAC_{400}, and MAC_{410}) for BrC formation during the photolysis of 3-methylcatechol in aqueous $NaNO_2$ solutions (pH 4–5, T = 25 °C) at different initial concentrations of $NaNO_2$ under simulated sunlight (YES) and in the dark (NO). The concentration of 3MC was always 0.1 mM.

Experiment	$NaNO_2$ (mM)	Sunlight	$MAC_{BrC(350nm)}$ (m^2/g)	$MAC_{BrC(400nm)}$ (m^2/g)	$MAC_{BrC(400nm)}$ (m^2/g)
1	0.1	YES	1.80	0.87	0.43
2	0.5	YES	3.10	1.26	0.94
3	1	YES	3.44	1.34	1.02
4	2	YES	3.77	1.54	1.17
5	5	YES	4.09	1.74	1.42
6	1	NO	5.41	1.73	1.75
7	2	NO	5.63	1.90	1.38

MAC values for the photolysis of 3MC in aqueous $NaNO_2$ solutions (from ca. 0.4 $m^2{\cdot}g^{-1}$ at 410 nm for 0.1 mM NO_2^- to ca. 4 $m^2{\cdot}g^{-1}$ at 350 nm for 5 mM NO_2^-, see Table 3) are of the same order of magnitude as reported in the literature [42]. Liu et al. [43] found out that MAC of SOA, especially those produced from aromatic precursors, are much higher for high-NO_x conditions than for low NO_x. In high-NO_x, SOA from toluene was largely composed of nitroaromatic compounds (e.g., NC, dinitrocatechol, and NP), the total absorbance of which accounted for 60% and 41% of the overall absorbance in the ranges of 300–400 and 400–500 nm, respectively [44]. Jiang et al. [45] has recently reported on MAC values of SOA from unsaturated heterocyclic organic compounds; for example, for pyrrole SOA from nighttime NO_3 oxidation, the value of $(MAC)_{290–700nm}$ was 0.34 ± 0.07 $m^2{\cdot}g^{-1}$. In addition, MAC values of water-soluble organic aerosols from different environments have been reported [42]; MAC of 0.3 $m^2{\cdot}g^{-1}$ at 532 nm was found for water-soluble Humic-like-substances(HULIS) fraction of BB aerosol (BBA) in Amason basin [46] and 0.5–1.5 $m^2{\cdot}g^{-1}$ at 404 nm for ambient BB organic aerosol (BBOA) in Boulder, Colorado [47]. Based on our results, we can conclude that during the reaction of 3MC and $NaNO_2$, both in the sunlight and in the dark, light-absorbing components are formed, largely MNC, which efficiently absorb light in the range characteristic of BrC. For comparison, in the aqueous reaction of 3MC at low $NaNO_2$ concentration (experiment 1), for a factor of more than 2 more BrC is formed (MAC is ca. 1.8 $m^2{\cdot}g^{-1}$ at 350 nm) than from toluene at high-NO_x conditions at moderate RH (MAC, 0.8 $m^2{\cdot}g^{-1}$ at 365 nm) [43].

4. Conclusions

This study demonstrates that aqueous-phase transformations of 3MC in the presence of HNO_2/NO_2^- under atmospheric sunlight conditions importantly contribute to the secondary BrC formation. The two primary products, 3M5NC and 3M4NC, which have been identified as the main products of nighttime processing and shown to strongly absorb in the near UV range [27], are also formed under the sunlight. However, their cumulative yield is only up to about 45% under illumination and was additionally found to be dependent on the initial concentration of NO_2^-. When NO_2^- is in excess with respect to 3MC, the rate constant of 3MC degradation (k_{2nd}(global) = 0.075 $M^{-1}{\cdot}s^{-1}$) is twice as high as in the dark (k_{2nd} = 0.032 $M^{-1}{\cdot}s^{-1}$) under the same initial conditions. Several different degradation mechanisms are possible in the sunlight. Besides the non-radical nitration reaction, which has already been studied in the dark, different reactive species can be formed during the photolysis of HNO_2 (e.g., OH, NO, NO_2), which are also possible sinks of 3MC.

The dependence of absorbance evolution on the initial NO_2^-/3MC concentration ratio during the photolysis showed that when the ratio is 1, the absorption around 350 nm increases with time, which corresponds to the formation of identified first-generation products (3M5NC and 3M4NC). At higher concentration ratios, the absorption at this wavelength substantially increases, and the absorption above 400 nm becomes pronounced as well. The second-generation products, i.e., 3M5NC-OH and the oxidative cleavage products of 3M4NC, were confirmed to be responsible for the absorption in the visible region.

With higher NO_2^- concentrations, the characteristic MAC values (up to more than 4 $m^2{\cdot}g^{-1}$ at 350 nm for 5 mM NO_2^-) increase due to more BrC formation. We believe that the aqueous-phase reaction of 3MC in the presence of HNO_2/NO_2^-, either under the sunlight or in the dark, may significantly contribute to SOA light absorption. This is supported by the fact that MNC and secondary hydroxylated/ring cleavage products produced from 3MC significantly absorb light in the range characteristic of BrC. Our conclusions are also supported by recent field measurements of aerosol particles, where MNC have been identified as significant contributors to atmospheric BrC [15–19,22].

Author Contributions: K.V. performed the experiments, conducted the research, and helped to write the manuscript; M.Š. helped in product identification; A.K. contributed to the data analysis and writing process; and I.G. wrote the manuscript. All authors have read and agreed to the published version of the manuscript.

Funding: This research received no external funding.

Acknowledgments: We acknowledge the financial support from the Slovenian Research Agency (research core funding No. P1-0034). We thank Luka Nunar for his help with experimental work.

Conflicts of Interest: The authors declare no competing financial interests.

References

1. Laskin, A.; Laskin, J.; Nizkorodov, S.A. Chemistry of Atmospheric Brown Carbon. *Chem. Rev.* **2015**, *115*, 4335–4382. [CrossRef]
2. Teich, M.; van Pinxteren, D.; Wang, M.; Kecorius, S.; Wang, Z.B.; Muller, T.; Mocnik, G.; Herrmann, H. Contributions of nitrated aromatic compounds to the light absorption of water-soluble and particulate brown carbon in different atmospheric environments in Germany and China. *Atmos. Chem. Phys.* **2017**, *17*, 1653–1672. [CrossRef]
3. Xie, M.J.; Chen, X.; Hays, M.D.; Holder, A.L. Composition and light absorption of N-containing aromatic compounds in organic aerosols from laboratory biomass burning. *Atmos. Chem. Phys.* **2019**, *19*, 2899–2915. [CrossRef]
4. Bond, T.C. Spectral dependence of visible light absorption by carbonaceous particles emitted from coal combustion. *Geophys. Res. Lett.* **2001**, *28*, 4075–4078. [CrossRef]
5. Feng, Y.; Ramanathan, V.; Kotamarthi, V.R. Brown carbon: A significant atmospheric absorber of solar radiation? *Atmos. Chem. Phys.* **2013**, *13*, 8607–8621. [CrossRef]
6. Andreae, M.O.; Gelencser, A. Black carbon or brown carbon? The nature of light-absorbing carbonaceous aerosols. *Atmos. Chem. Phys.* **2006**, *6*, 3131–3148. [CrossRef]
7. Saleh, R.; Robinson, E.S.; Tkacik, D.S.; Ahern, A.T.; Liu, S.; Aiken, A.C.; Sullivan, R.C.; Presto, A.A.; Dubey, M.K.; Yokelson, R.J. Brownness of organics in aerosols from biomass burning linked to their black carbon content. *Nat. Geosci.* **2014**, *7*, 647–650. [CrossRef]
8. Pflieger, M.; Kroflič, A. Acute toxicity of emerging atmospheric pollutants from wood lignin due to biomass burning. *J. Hazard. Mater.* **2017**, *338*, 132–139. [CrossRef]
9. Wang, L.; Wang, X.; Gu, R.; Wang, H.; Yao, L.; Wen, L.; Zhu, F.; Wang, W.; Xue, L.; Yang, L. Observations of fine particulate nitrated phenols in four sites in northern China: concentrations, source apportionment, and secondary formation. *Atmos. Chem. Phys.* **2018**, *18*, 4349–4359. [CrossRef]
10. Saleh, R.; Hennigan, C.J.; McMeeking, G.R.; Chuang, W.K.; Robinson, E.S.; Coe, H.; Donahue, N.M.; Robinson, A.L. Absorptivity of brown carbon in fresh and photo-chemically aged biomass-burning emissions. *Atmos. Chem. Phys.* **2013**, *13*, 7683–7693. [CrossRef]
11. Kroflič, A.; Grilc, M.; Grgić, I. Unraveling Pathways of Guaiacol Nitration in Atmospheric Waters: Nitrite, A Source of Reactive Nitronium Ion in the Atmosphere. *Environ. Sci. Technol.* **2015**, *49*, 9150–9158. [CrossRef]
12. McNeill, V.F. Aqueous Organic Chemistry in the Atmosphere: Sources and Chemical Processing of Organic Aerosols. *Environ. Sci. Technol.* **2015**, *49*, 1237–1244. [CrossRef]
13. Harrison, M.A.J.; Barra, S.; Borghesi, D.; Vione, D.; Arsene, C.; Olariu, R.L. Nitrated phenols in the atmosphere: a review. *Atmos. Environ.* **2005**, *39*, 231–248. [CrossRef]
14. Hoffmann, E.H.; Tilgner, A.; Wolke, R.; Boge, O.; Walter, A.; Herrmann, H. Oxidation of substituted aromatic hydrocarbons in the tropospheric aqueous phase: Kinetic mechanism development and modelling. *Phys. Chem. Chem. Phys.* **2018**, *20*, 10960–10977. [CrossRef]
15. Claeys, M.; Vermeylen, R.; Yasmeen, F.; Gomez-Gonzalez, Y.; Chi, X.G.; Maenhaut, W.; Meszaros, T.; Salma, I. Chemical characterisation of humic-like substances from urban, rural and tropical biomass burning environments using liquid chromatography with UV/vis photodiode array detection and electrospray ionisation mass spectrometry. *Environ. Chem.* **2012**, *9*, 273–284. [CrossRef]
16. Kitanovski, Z.; Grgić, I.; Vermeylen, R.; Claeys, M.; Maenhaut, W. Liquid chromatography tandem mass spectrometry method for characterization of monoaromatic nitro-compounds in atmospheric particulate matter. *J. Chromatogr. A* **2012**, *1268*, 35–43. [CrossRef]
17. Kahnt, A.; Behrouzi, S.; Vermeylen, R.; Shalamzari, M.S.; Vercauteren, J.; Roekens, E.; Claeys, M.; Maenhaut, W. One-year study of nitro-organic compounds and their relation to wood burning in PM10 aerosol from a rural site in Belgium. *Atmos. Environ.* **2013**, *81*, 561–568. [CrossRef]
18. Chow, K.S.; Huang, X.H.H.; Yu, J.Z. Quantification of nitroaromatic compounds in atmospheric fine particulate matter in Hong Kong over 3 years: Field measurement evidence for secondary formation derived from biomass burning emissions. *Environ. Chem.* **2016**, *13*, 665–673. [CrossRef]
19. Wang, X.F.; Gu, R.R.; Wang, L.W.; Xu, W.X.; Zhang, Y.T.; Chen, B.; Li, W.J.; Xue, L.K.; Chen, J.M.; Wang, W.X. Emissions of fine particulate nitrated phenols from the burning of five common types of biomass. *Environ. Pollut.* **2017**, *230*, 405–412. [CrossRef]

20. Pereira, K.L.; Hamilton, J.F.; Rickard, A.R.; Bloss, W.J.; Alam, M.S.; Camredon, M.; Ward, M.W.; Wyche, K.P.; Munoz, A.; Vera, T.; et al. Insights into the Formation and Evolution of Individual Compounds in the Particulate Phase during Aromatic Photo-Oxidation. *Environ. Sci. Technol.* **2015**, *49*, 13168–13178. [CrossRef]
21. Yuan, B.; Liggio, J.; Wentzell, J.; Li, S.-M.; Stark, H.; Roberts, J.M.; Gilman, J.; Lerner, B.; Warneke, C.; Li, R. Secondary formation of nitrated phenols: Insights from observations during the Uintah Basin Winter Ozone Study (UBWOS) 2014. *Atmos. Chem. Phys.* **2016**, *16*, 2139–2153. [CrossRef]
22. Frka, S.; Šala, M.; Kroflič, A.; Huš, M.; Čusak, A.; Grgić, I. Quantum Chemical Calculations Resolved Identification of Methylnitrocatechols in Atmospheric Aerosols. *Environ. Sci. Technol.* **2016**, *50*, 5526–5535. [CrossRef] [PubMed]
23. Kroflič, A.; Huš, M.; Grilc, M.; Grgić, I. Underappreciated and Complex Role of Nitrous Acid in Aromatic Nitration under Mild Environmental Conditions: The Case of Activated Methoxyphenols. *Environ. Sci. Technol.* **2018**, *52*, 13756–13765. [CrossRef] [PubMed]
24. Lim, Y.B.; Tan, Y.; Perri, M.J.; Seitzinger, S.P.; Turpin, B.J. Aqueous chemistry and its role in secondary organic aerosol (SOA) formation. *Atmos. Chem. Phys.* **2010**, *10*, 10521–10539. [CrossRef]
25. Ervens, B.; Turpin, B.J.; Weber, R.J. Secondary organic aerosol formation in cloud droplets and aqueous particles (aqSOA): A review of laboratory, field and model studies. *Atmos. Chem. Phys.* **2011**, *11*, 11069–11102. [CrossRef]
26. Iinuma, Y.; Boge, O.; Grafe, R.; Herrmann, H. Methyl-Nitrocatechols: Atmospheric Tracer Compounds for Biomass Burning Secondary Organic Aerosols. *Environ. Sci. Technol.* **2010**, *44*, 8453–8459. [CrossRef]
27. Vidović, K.; Jurković, D.L.; Šala, M.; Kroflič, A.; Grgić, I. Nighttime Aqueous-Phase Formation of Nitrocatechols in the Atmospheric Condensed Phase. *Environ. Sci. Technol.* **2018**, *52*, 9722–9730. [CrossRef]
28. Vidović, K.; Kroflič, A.; Jovanovič, P.; Šala, M.; Grgić, I. Electrochemistry as a Tool for Studies of Complex Reaction Mechanisms: The Case of the Atmospheric Aqueous-Phase Aging of Catechols. *Environ. Sci. Technol.* **2019**, *53*, 11195–11203. [CrossRef]
29. Albinet, A.; Minero, C.; Vione, D. UVA irradiation induces direct phototransformation of 2, 4-dinitrophenol in surface water samples. *Chemosphere* **2010**, *80*, 759–763. [CrossRef]
30. Zhao, S.F.; Ma, H.J.; Wang, M.; Cao, C.Q.; Xiong, J.; Xu, Y.S.; Yao, S.D. Study on the mechanism of photo-degradation of p-nitrophenol exposed to 254 nm UV light. *J. Hazard. Mater.* **2010**, *180*, 86–90. [CrossRef]
31. Slade, J.H.; Knopf, D.A. Heterogeneous OH oxidation of biomass burning organic aerosol surrogate compounds: Assessment of volatilisation products and the role of OH concentration on the reactive uptake kinetics. *Phys. Chem. Chem. Phys.* **2013**, *15*, 5898–5915. [CrossRef] [PubMed]
32. Zhao, R.; Lee, A.K.Y.; Huang, L.; Li, X.; Yang, F.; Abbatt, J.P.D. Photochemical processing of aqueous atmospheric brown carbon. *Atmos. Chem. Phys.* **2015**, *15*, 6087–6100. [CrossRef]
33. Hems, R.F.; Abbatt, J.P.D. Aqueous Phase Photo-oxidation of Brown Carbon Nitrophenols: Reaction Kinetics, Mechanism, and Evolution of Light Absorption. *ACS Earth Space Chem.* **2018**, *2*, 225–234. [CrossRef]
34. Palumbo, A.; Napolitano, A.; d'Ischia, M. Nitrocatechols versus nitrocatecholamines as novel competitive inhibitors of neuronal nitric oxide synthase: Lack of the aminoethyl side chain determines loss of tetrahydrobiopterin-antagonizing properties. *Bioorg. Med. Chem. Lett.* **2002**, *12*, 13–16. [CrossRef]
35. Kitanovski, Z.; Grgić, I.; Yasmeen, F.; Claeys, M.; Čusak, A. Development of a liquid chromatographic method based on ultraviolet-visible and electrospray ionization mass spectrometric detection for the identification of nitrocatechols and related tracers in biomass burning atmospheric organic aerosol. *Rapid Commun. Mass Spectrom.* **2012**, *26*, 793–804. [CrossRef]
36. Kitanovski, Z.; Čusak, A.; Grgić, I.; Claeys, M. Chemical characterization of the main products formed through aqueous-phase photonitration of guaiacol. *Atmos. Meas. Tech.* **2014**, *7*, 2457–2470. [CrossRef]
37. Vione, D.; Maurino, V.; Minero, C.; Pelizzetti, E. Phenol photonitration upon UV irradiation of nitrite in aqueous solution I: Effects of oxygen and 2-propanol. *Chemosphere* **2001**, *45*, 893–902. [CrossRef]
38. He, L.; Schaefer, T.; Otto, T.; Kroflič, A.; Herrmann, H. Kinetic and Theoretical Study of the Atmospheric Aqueous-Phase Reactions of OH Radicals with Methoxyphenolic Compounds. *J. Phys. Chem. A* **2019**, *123*, 7828–7838. [CrossRef]
39. Kroflič, A.; Schaefer, T.; Huš, M.; Phuoc Le, H.; Otto, T.; Herrmann, H. OH radicals reactivity towards phenol-related pollutants in water: temperature dependence of the rate constants and novel insights into the [OH–phenol]$^{\cdot}$ adduct formation. *Phys. Chem. Chem. Phys.* **2020**, *22*, 1324–1332. [CrossRef]

40. Nakayama, T.; Sato, K.; Matsumi, Y.; Imamura, T.; Yamazaki, A.; Uchiyama, A. Wavelength and NO_x dependent complex refractive index of SOAs generated from the photooxidation of toluene. *Atmos. Chem. Phys.* **2013**, *13*, 531–545. [CrossRef]
41. Li, Y.J.; Huang, D.D.; Cheung, H.Y.; Lee, A.K.Y.; Chan, C.K. Aqueous-phase photochemical oxidation and direct photolysis of vanillin—A model compound of methoxy phenols from biomass burning. *Atmos. Chem. Phys.* **2014**, *14*, 2871–2885. [CrossRef]
42. Bluvshtein, N.; Lin, P.; Flores, J.M.; Segev, L.; Mazar, Y.; Tas, E.; Snider, G.; Weagle, C.; Brown, S.S.; Laskin, A.; et al. Broadband optical properties of biomass-burning aerosol and identification of brown carbon chromophores. *J. Geophys. Res. Atmos.* **2017**, *122*, 5441–5456. [CrossRef]
43. Liu, J.M.; Lin, P.; Laskin, A.; Laskin, J.; Kathmann, S.M.; Wise, M.; Caylor, R.; Imholt, F.; Selimovic, V.; Shilling, J.E. Optical properties and aging of light-absorbing secondary organic aerosol. *Atmos. Chem. Phys.* **2016**, *16*, 12815–12827. [CrossRef]
44. Lin, P.; Liu, J.M.; Shilling, J.E.; Kathmann, S.M.; Laskin, J.; Laskin, A. Molecular characterization of brown carbon (BrC) chromophores in secondary organic aerosol generated from photo-oxidation of toluene. *Phys. Chem. Chem. Phys.* **2015**, *17*, 23312–23325. [CrossRef]
45. Jiang, H.H.; Frie, A.L.; Lavi, A.; Chen, J.Y.; Zhang, H.F.; Bahreini, R.; Lin, Y.H. Brown Carbon Formation from Nighttime Chemistry of Unsaturated Heterocyclic Volatile Organic Compounds. *Environ. Sci. Technol. Lett.* **2019**, *6*, 184–190. [CrossRef]
46. Hoffer, A.; Gelencser, A.; Guyon, P.; Kiss, G.; Schmid, O.; Frank, G.P.; Artaxo, P.; Andreae, M.O. Optical properties of humic-like substances (HULIS) in biomass-burning aerosols. *Atmos. Chem. Phys.* **2006**, *6*, 3563–3570. [CrossRef]
47. Lack, D.A.; Richardson, M.S.; Law, D.; Langridge, J.M.; Cappa, C.D.; McLaughlin, R.J.; Murphy, D.M. Aircraft Instrument for Comprehensive Characterization of Aerosol Optical Properties, Part 2: Black and Brown Carbon Absorption and Absorption Enhancement Measured with Photo Acoustic Spectroscopy. *Aerosol Sci. Technol.* **2012**, *46*, 555–568. [CrossRef]

atmosphere

MDPI

Article

Aqueous Reactions of Sulfate Radical-Anions with Nitrophenols in Atmospheric Context

Krzysztof J. Rudziński * and **Rafał Szmigielski**

Institute of Physical Chemistry of the PAS, Kasprzaka 44/52, 01-224 Warsaw, Poland; rafal.szmigielski@gmail.com

* Correspondence: kjrudz@ichf.edu.pl; Tel.: +48-22-343-3402

Received: 14 November 2019; Accepted: 4 December 2019; Published: 9 December 2019

Abstract: Nitrophenols, hazardous environmental pollutants, react promptly with atmospheric oxidants such as hydroxyl or nitrate radicals. This work aimed to estimate how fast nitrophenols are removed from the atmosphere by the aqueous-phase reactions with sulfate radical-anions. The reversed-rates method was applied to determine the relative rate constants for reactions of 2-nitrophenol, 3-nitrophenol, 4-nitrophenol, 2,4-dinitrophenol, and 2,4,6-trinitrophenol with sulfate radical-anions generated by the autoxidation of sodium sulfite catalyzed by iron(III) cations at ~298 K. The constants determined were: 9.08×10^8, 1.72×10^9, 6.60×10^8, 2.86×10^8, and 7.10×10^7 M^{-1} s^{-1}, respectively. These values correlated linearly with the sums of Brown substituent coefficients and with the relative strength of the O–H bond of the respective nitrophenols. Rough estimation showed that the gas-phase reactions of 2-nitrophenol with hydroxyl or nitrate radicals dominated over the aqueous-phase reaction with sulfate radical-anions in deliquescent aerosol and haze water. In clouds, rains, and haze water, the aqueous-phase reaction of 2-nitrophenol with sulfate radical-anions dominated, provided the concentration of the radical-anions was not smaller than that of the hydroxyl or nitrate radicals. The results presented may be also interesting for designers of advanced oxidation processes for the removal of nitrophenol.

Keywords: atmospheric processes; secondary organic aerosol; rate constants; atmospheric pollutants; advanced oxidation processes

1. Introduction

Nitrophenols are well known environmental trace compounds and pollutants [1,2], which have been detected in various environmental matrices including air [3–6], rainwater [3,7–9], cloud water [10], fog [10,11], snow [1], atmospheric aerosol [4–6,12–25], soils [26,27], and surface waters [10,28–32]. They originate from many anthropogenic and natural sources including: the incineration of wastes [33], industrial chemical processes [34], combustion of coal and biomass as well as vehicle and aviation fuels [35–37], degradation of pesticides [34,38,39], release of wood preservatives [34], and atmospheric chemical reactions. 4-nitrophenol (4-NP) and 2,4-dinitrophenol (2,4-DNP), along with sugar anhydrides such as levoglucosan, serve as markers of biomass burning in ambient aerosol [40–43]. These compounds and 2-nitrophenol are recognized components of atmospheric brown carbon, i.e., a collection of light absorbing organic compounds in the atmosphere [22,44,45]

Atmospheric reactions that yield nitrophenols take place both in the gas phase and in the aqueous phase. For instance, the gas-phase nitration of phenol involves hydroxyl radicals $^{\bullet}OH$ and NO_2 in the daytime or nitrate radical $^{\bullet}NO_3$ and NO_2 in the night to respectively produce 2-nitrophenol (2-NP) or 2-NP and 4-NP. The chemical mechanisms of both processes were thoroughly reviewed [1]. Formation of nitrophenols in atmospheric waters of all kinds is at least equally important but less understood. The possible pathways include oxidation of phenols with NO_2 and OH or NO_3 radicals [46], electrophilic nitration initiated by N_2O_5 and $ClNO_2$ [1,47], photolytic and dark reactions involving nitrate radicals, inorganic nitrates, nitrites and nitrous acid HONO [48,49], and photolytic reactions with nitrogen dioxide in the presence of iron oxide and oxygen [49].

Atmospheric sinks for nitrophenols include photolysis [50] and the gas-phase reactions with OH radicals and NO_3 radicals [51,52] which are characterized by the estimated residence time of several days. More efficient sinks may include partitioning to atmospheric aqueous phases followed by reactions with various radicals and/or photolysis [1,46]. More recently, Barsotti, et al. [53] demonstrated that the irradiation of aqueous solutions or viscous films containing several nitrophenols (2-NP, 4-NP, 2,4-DNP, and 2,6-DNP, i.e., 2,6-dinitrophenol) was an efficient source of HONO and NO_2^- ions. Vione, et al. [54] showed that OH radicals reacted faster than NO_3 radicals with 2-NP and 4-NP in aqueous solutions to lower the atmospheric levels of 2-NP below those of 4-NP. Hems and Abbatt [55] studied the aqueous-phase photo-oxidation of 2,4-DNP by OH radicals, identified numerous intermediate products thereof and showed the corresponding evolution of UV absorbance of the reacting solutions. In addition, many laboratories studied the aqueous-phase reactions of other substituted phenols of atmospheric interest like guaiacol, nitro-guaiacol, vanillin, or syringol [55–59].

Much of the nitrophenol chemistry has been studied for the sake of advanced oxidation processes aimed at mitigation of nitrophenols in aquatic and industrial environments [60]. The technologies considered include: Fenton and photo-Fenton reactions based on H_2O_2 [34,61], TiO_2 based photocatalysis [62–64], electrocatalysis [65], photo-electrocatalysis [66,67], and wet catalysis [68,69]. Among the latter, a promising process was proposed which utilized reactions of nitrophenols with sulfate radical-anions generated by the cobalt-mediated decomposition of peroxymonosulfate anions [70].

For years, some nitrophenols (2-NP, 4-NP, 2,4-DNP) have been listed as priority or hazardous pollutants [71–73]. Generally, mono- and di-nitrophenols are considered toxic in plants and mammals [74], while 4-nitrophenol is highly toxic in humans [75]. Although EPA USA has not considered 4-nitrophenol carcinogenic [38], a laboratory experiment showed the compound can destroy DNA in vitro [76].

This work was aimed at elucidating how fast nitrophenols are removed from the atmospheric waters by reaction with sulfate radical-anions, which are important atmospheric oxidants known to react fast with numerous atmospheric pollutants [77–83].

2. Experiments

2.1. Chemicals

The following chemicals were used as purchased: 2-nitrophenol (R.G.), 3-nitrophenol (REAGENTPLUS™, 99%), 4-nitrophenol and 2,4,6-trinitrophenol (1 wt % solution in water) from Sigma Aldrich, 2,4-dinitrophenol (97% + 15% H_2O) from Alfa Aesar, $Fe(ClO_4)_3{\cdot}9H_2O$ (purum) from Fluka, $Na_2S_2O_5$ (EMSURE® ACS, Reag. Ph Eur. > 98%) and $HClO_4$ (pro analysis) from Merck, argon (99.999%) from Multax. For each experiment, aqueous solutions of reactants were prepared freshly using Milli-Q water (18.2 MΩ cm, Milli-Q Advantage System from Merck Millipore). Buffer standards used for the calibration of pH electrodes were from Thermo Fisher Scientific. To avoid the contact with the atmospheric oxygen, Milli-Q water was deoxygenated by a stream of argon bubbled through for 20 min. Solutions of sodium bisulfite were obtained by dissolving $Na_2S_2O_5$ in deoxygenated Milli-Q water

$$Na_2S_2O_5 + H_2O \rightleftarrows 2Na^+ + 2HSO_3^-, \quad (1)$$

The acidity of solutions was adjusted to pH = 3.1 with 0.1 M $HClO_4$ so the species in solutions were predominantly Na^+ and HSO_3^- ions.

2.2. Estimation of the Rate Constants

The relative rate constants for reactions of nitrophenols with sulfate radical-anions were estimated using the reversed-rates method developed by Ziajka and Pasiuk-Bronikowska [84] and successfully applied to several organic compounds [83,85,86]. Briefly, the sulfate radical-anions are generated during chain autoxidation of sulfite anions catalyzed by Fe(III) cations. The mechanism of the autoxidation was presented in detail by Ziajka and Rudziński [83] and recalled in the SI. One runs several experiments with the autoxidation of S(IV) inhibited by two different compounds, inh1 and inh2, used at several different initial concentrations (e.g., Figure 1). Each experiment should attain a pseudo-stationary phase during which the autoxidation proceeds at a constant rate (e.g., Figure 2). Then, one plots the reciprocal stationary rates observed against the initial concentrations of the inhibitor used (Figure 3). If both plots are linear, the ratio of rate constants for reactions of the inhibitors with sulfate radical-anions is equal to the ratio of the slopes of the linear plots (Equation (2)). If one knows the rate constant for reaction of one inhibitor with sulfate radical-anions, one can calculate the rate constant for the other inhibitor.

$$k_{\text{inh1}+\text{SO}_4^-} = \frac{\text{slope}_{\text{inh1}}}{\text{slope}_{\text{inh2}}} k_{\text{inh2}+\text{SO}_4^-}, \quad (2)$$

In the present work, ethanol was used as a reference inhibitor against which the rate constants for nitrophenols were calculated.

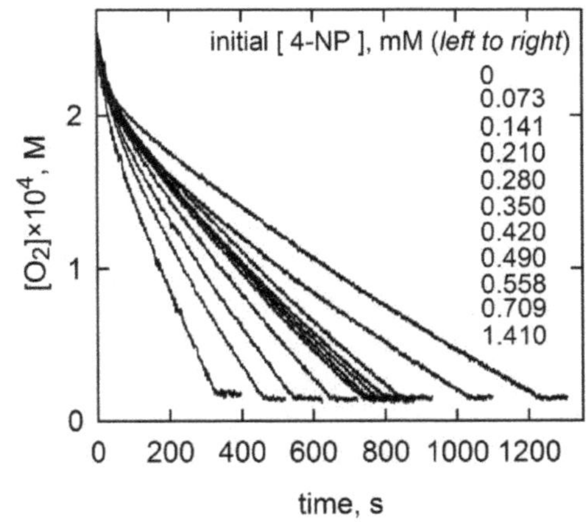

Figure 1. Concentration of oxygen recorded during autoxidation of $NaHSO_3$ inhibited by 4-NP at various initial concentrations.

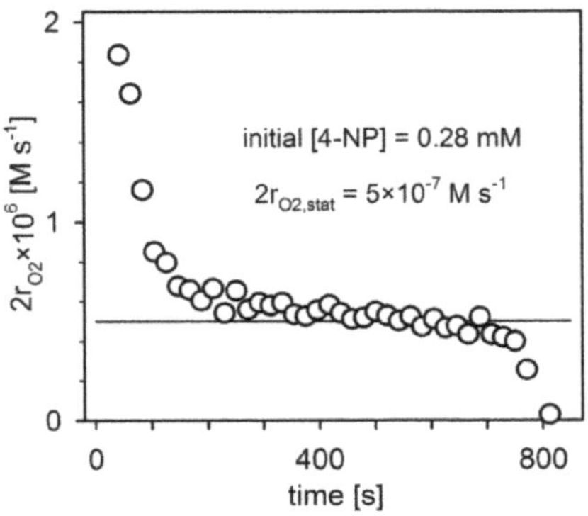

Figure 2. Rate of oxygen consumption during autoxidation of $NaHSO_3$ inhibited by 4-NP (initially 0.28 mM) evaluated from data in Figure 1.

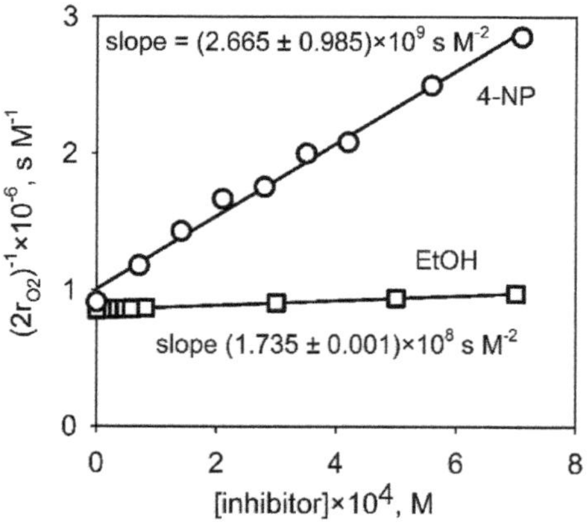

Figure 3. Linear plots of reciprocal quasi-stationary rates of the autoxidation of $NaHSO_3$ inhibited by 4-NP or by the reference ethanol versus initial concentrations of each inhibitor. The slope uncertainties are equal to standard errors of regression coefficients.

2.3. Experimental Runs

The experimental setup and procedure for carrying out the stationary autoxidation of S(VI) inhibited by organic compounds was described in detail elsewhere [83]. Briefly, the experiments were carried out in a well-mixed glass reactor of 60 cm^3 volume, closed with a Teflon cover and thermostatted at 298 K within a water jacket. For each run, the reactor was filled with aqueous solution of sodium bisulfite and oxygen so that it contained no gas phase. The pH of solution was adjusted to 3.1 with $HClO_4$. The pH of the solution was recorded using a SenTix Mic combination pH electrode from WTW. Then, a small aliquot of aqueous solution of $Fe(ClO_4)_3$ catalyst was injected to start the reaction. Table 1 shows the initial concentrations of reactants. The autoxidation of S(IV) was followed by recording the concentration of oxygen using an Orion 97-08 from Thermo Fisher Scientific and a home-designed pH/oxygen meter and software. Equation (3a) shows the overall stoichiometry of uninhibited SIV autoxidation. Assuming the conversion of inhibitors was small, the stoichiometry of the autoxidation inhibited by a nitrophenol was defined by the same equation so that the rate of the autoxidation was defined by Equation (3b) [83].

$$2SO_3^{2-}/2HSO_3^- + O_2 = SO_4^{2-}/HSO_4^-, \tag{3a}$$

$$r_{autoxidation} = -\frac{d[\mathrm{S(IV)}]}{dt} = \frac{d[\mathrm{S(VI)}]}{dt} = -2\frac{d[\mathrm{O_2}]}{dt}, \tag{3b}$$

Table 1. Initial concentrations of reactants used in the experiments.

Compound	Abbreviation	Concentration Range, mM
2-nitrophenol	2-NP	0.032–0.347
3-nitrophenol	3-NP	0.019–0.089
4-nitrophenol	4-NP	0.073–0.711
2,4-dinitrophenol	2,4-DNP	0.169–1.325
2,4,6-trinitrophenol	2,4,6-TNP	0.197–0.788
HSO_3^-		2
O_2		~0.25
$Fe(ClO_4)_3$		0.01

2.4. *Correction of the Diffusional Limitations of the Rate Constants*

Since the reactions examined were very fast, we corrected the rate constants determined for diffusional limitations using a simple resistance-in-series model [87–90]

$$k_{observed}^{-1} = k_{reaction}^{-1} + k_{diffusion}^{-1}, \tag{4}$$

$$k_{diffusion} = 4\pi(D_A + D_B)(r_A + r_B)N \times 10^3, \tag{5}$$

where all k are second order rate constants (M^{-1} s^{-1}), D are diffusion coefficients of reactants A and B (m^2 s^{-1}), r are reaction radii of reactant molecules A and B (m), and N is the Avogadro number (mol^{-1}). Details of the calculations are summarized in Section S3 of the SI.

3. Results

In this section, we present the experimental results obtained for 4-NP. The results for other nitrophenols were similar so we present them in the SI. Figure 1 shows consumption of oxygen during the autoxidation of S(IV) in the presence of 4-NP. The higher was the initial concentration of nitrophenol, the slower was the consumption of O_2.

In each experiment, the autoxidation attained a quasi-stationary rate, as shown in Figure 2 for the run with [4-NP] = 0.28 mM.

Figure 3 shows the plots of reciprocal stationary rates for autoxidation of S(IV) in the presence of 4-NP or a reference compound ethanol versus initial concentrations of each inhibitor. The plots were linear, so their slopes were used in Equation (2) to calculate the relative rate constant for the reaction of 4-NP with sulfate radical-anions

$$k_{4-\mathrm{NP}+\mathrm{SO}_4^-} = k_{\mathrm{EtOH}+\mathrm{SO}_4^-} \frac{\mathrm{slope}_{4-\mathrm{NP}}}{\mathrm{slope}_{\mathrm{EtOH}}} = 4.3 \times 10^7 \frac{2.665 \times 10^9}{1.735 \times 10^8} = 6.636 \times 10^8 \ \mathrm{M}^{-1}\mathrm{s}^{-1}, \tag{6}$$

Plots for other nitrophenols, all of them linear, were placed in the SI. The results of all experiments are collected in Table 2 and include the slopes of linear plots and the rate constants for reactions of nitrophenols with sulfate radical-anions, both observed and corrected for diffusional limitations. The uncertainties of the observed rate constants were estimated using the total differential method applied to Equation (6) with individual errors equal to the standard errors of the linear slopes and k_{EtOH} (Table 2). The uncertainties of the corrected rate constants were estimated in a similar way from Equation (4), assuming arbitrarily the uncertainty of $k_{difffusion}$ was 10%.

Table 2. Experimental slopes of linear plots (Figure 3 and Figure S2) and rate constants for reactions of nitrophenols with sulfate radical-anions (observed and corrected for diffusional limitations).

Compound	Slope, s M^{-2}	$k_{observed}$, $M^{-1}s^{-1}$	$k_{diffusion}$ [A], $M^{-1}s^{-1}$	$k_{reaction}$, $M^{-1}s^{-1}$	$(k_r - k_o)/k_o$, %
EtOH (*reference*)	$(1.735 \pm 0.001) \times 10^8$			$(4.30 \pm 0.86) \times 10^7$	
2-NP	$(3.662 \pm 0.235) \times 10^9$	$(9.08 \pm 2.40) \times 10^8$	2.01×10^{10}	$(9.50 \pm 4.58) \times 10^8$	4.74
3-NP	$(6.957 \pm 0.277) \times 10^9$	$(1.72 \pm 0.41) \times 10^9$	2.01×10^{10}	$(1.89 \pm 0{,}89) \times 10^9$	9.53
4-NP	$(2.665 \pm 0.985) \times 10^9$	$(6.60 \pm 3.77) \times 10^8$	2.01×10^{10}	$(6.83 \pm 5.42) \times 10^8$	3.41
2,4-DNP	$(1.154 \pm 0.152) \times 10^9$	$(2.86 \pm 0.95) \times 10^8$	2.13×10^{10}	$(2.90 \pm 1.58) \times 10^8$	1.51
2,4,6-TNP	$(2.865 \pm 0.180) \times 10^8$	$(7.10 \pm 1.87) \times 10^7$	2.18×10^{10}	$(7.12 \pm 3.31) \times 10^7$	0.33

A—assumed unceratinty 2×10^9 M^{-1} s^{-1}(10%).

4. Discussion

4.1. *Hammett's Correlations*

The reactions of nitrophenols with sulfate radical-anions appeared quite fast, with observed second order rate constants of the order 10^9 s^{-1} M^{-1} (Table 2). However, the diffusional limitations were not significant because the rate constants corrected for diffusional limitations were higher for a few percent only. The rate constants decreased with the number of NO_2 groups in the molecule with the exception of 3-NP. Figure 4a shows the uncorrected rate constants for reactions of sulfate radical-anions with phenol and substituted phenols—chlorophenols [83] and nitrophenols (this work)—correlate well with sums of Brown substituent coefficients for the compounds. The Brown coefficients for chlorophenols were taken after [83], while those for nitrophenols were: $\sigma_m{}^+ = 0.71$, $\sigma_p{}^+ = 0.79$ [91,92] and $\sigma_o{}^+ = 0.66$ $\sigma_p{}^+ = 0.52$ [93] for *meta, para* and *orto* substituents, respectively. The straight line in Figure 4a was obtained by linear regression covering all data (Equation (7)).

$$\log\left(k_{SO_4}\right) = (9.9006 \pm 0.0785) - (1.1513 \pm 0.0970)\sum \sigma^+, \quad R^2 = 0.9663, \tag{7}$$

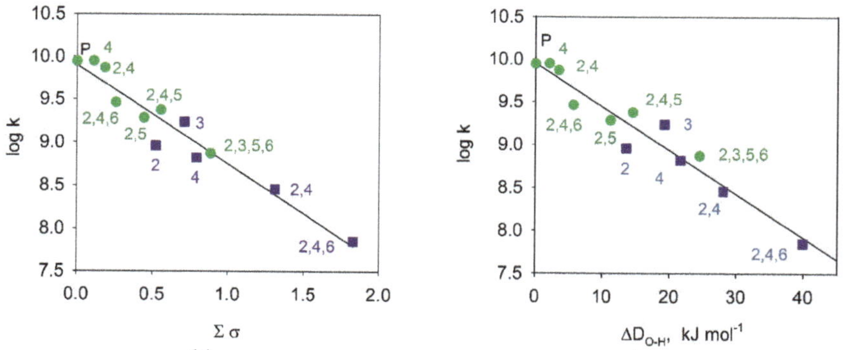

(**a**) (**b**)

Figure 4. Correlation of rate constants for reactions of sulfate radical-anions with phenol (P), chlorophenols (green circles: 4-CP; 2,4-DCP; 2,5-DCP; 2,4,5-TCP; 2,4,6-TCP; 2,3,5,6-TTCP) [83] and nitrophenols (blue squares: 2-NP; 3-NP; 4-NP; 2,4-DNP; 2,4,6-TNP) [this work] against (**a**) sums of Brown substituent coefficients and (**b**) the relative strength of the O–H bond.

Equation (7) can be used to estimate the second order rate constants for reactions of sulfate radical-anions with substituted phenols that had not been determined experimentally.

The Brown substituent coefficients can estimate the relative strength of the O–H bonds in substituted phenols with a linear correlation developed by Jonsson et al. [93] (Equation (8)).

$$\Delta D_{O\text{-}H} = -2 + 29.9\left(\sigma^+_{o2} + \sigma^+_{m3} + \sigma^+_{p4} + \sigma^+_{m5} + \sigma^+_{o6}\right), \quad \text{kJ mol}^{-1}, \tag{8}$$

Therefore, the rate constants for reactions of sulfate radical-anions with phenol and substituted phenols can also be correlated against $\Delta D_{O\text{-}H}$ (Figure 4b). The corresponding linear regression is given by Equation (9).

$$\log\left(k_{SO_4}\right) = (9.9498 \pm 0.0828) - (0.0508 \pm 0.0043)\Delta D_{O\text{-}H}, \quad R^2 = 0.9665, \tag{9}$$

Equation (9) can estimate the second order rate constants for reactions of sulfate radical-anions with substituted phenols in place of Equation (7).

4.2. Atmospheric Significance

The atmospheric significance of the aqueous-phase reactions of nitrophenols with sulfate radical-anions was evaluated using the approach developed in [81]. The rate of the total conversion of a nitrophenol NP by a reactant X (OH or NO_3) in the gas phase ($r_{X,g}$) and in the aqueous phase ($r_{X,aq} \times \omega$) was compared to the rate of conversion of this NP by sulfate radical-anions in the aqueous phase within the gas phase ($r_{SO_4,aq}\ \omega$) as the ratio $R_{X,tot\text{-}aq}$ defined by Equation (10). The concentrations of the gas-phase and aqueous-phase reactants were assumed to follow the Henry's Law.

$$R_{X,tot-aq} = \frac{r_{X,g} + r_{X,aq}\omega}{r_{SO_4,aq}\omega} = \frac{k_{X,g}[X]_g[NP]_g + k_{X,aq}[X]_{aq}[NP]_{aq}\omega}{k_{SO_4,aq}[X]_{aq}\left[SO_4^{\bullet-}\right]_{aq}\omega} = \frac{\frac{k_{X,g}}{H_{d,X}H_{d,NP}} + k_{X,aq}\omega}{k_{SO_4,aq}\omega} \cdot \frac{[X]_{aq}}{\left[SO_4^{\bullet-}\right]_{aq}}, \quad (10)$$

where ω m^3 m^{-3} is the atmospheric liquid water contents; $k_{X,g}$ and $k_{X,aq}$ dm^3 mol^{-1} s^{-1} are the rate constant for the reaction of X with NP in the gas phase and the aqueous phase, respectively; $k_{SO_4,aq}$ dm^3 mol^{-1} s^{-1} is the rate constant for the reaction of $SO_4^{\bullet-}$ with NP in the aqueous phase; H_d is the dimensionless Henry's constant ($H_d = H \times$ gas constant $\times$ absolute temperature) for X or for NP; $[X]_{aq}$ and $[SO_4^{\bullet-}]_{aq}$ are the aqueous-phase concentrations of X and $SO_4^{\bullet-}$.

Figure 5 compares the total conversion of 2-NP due to the gas-phase and the aqueous-phase reactions with OH radicals to the aqueous-phase reaction with $SO_4^{\bullet-}$ radical-anions (a) as well as the conversion of 2-NP due to the gas-phase and aqueous-phase reactions with NO_3 radicals to the aqueous-phase reaction with $SO_4^{\bullet-}$ radical-anions (b). Data required for the calculations behind the plots are given in Tables 2 and 3 while more details are available in the SI. The ranges of radical concentrations considered in Figure 5 fit well within the realistic ranges estimated by modeling for the atmospheric systems (Table 4) [94].

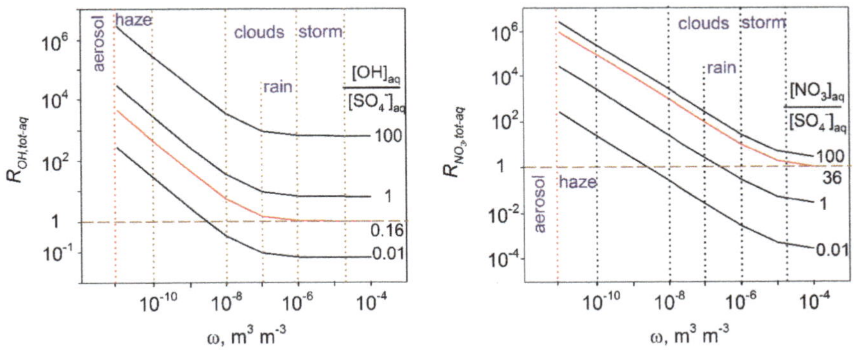

(a) (b)

Figure 5. Total rate of 2-NP conversion in the atmosphere due the gas-phase and aqueous-phase reactions with OH radicals (**a**) or NO_3 radicals (**b**) compared to the rate of the aqueous-phase reaction of 2-NP with sulfate radical-anions for various ratios of radicals in the aqueous phases ($[OH]_{aq}/[SO_4^{\bullet-}]_{aq}$) and various liquid water contents ω (based on Equation (10)).

In most cases, the total OH radical sink for 2-NP dominates over the $SO_4^{\bullet-}$ aqueous sink in all atmospheric aqueous phases. Sulfate radical-anions take the lead only if they are in significant excess in clouds, rains, and storms (red line in Figure 5a, $[OH]_{aq}/[SO_4^{\bullet-}]_{aq} < 0.16$). The total NO_3 radical sink for 2-NP dominates over the $SO_4^{\bullet-}$ aqueous sink in aerosol and haze waters and in clouds and rains if in significant excess (red line in Figure 5b, $[NO_3]_{aq}/[SO_4^{\bullet-}]_{aq} > 36$). In other cases, the $SO_4^{\bullet-}$ aqueous sink prevails. The above rationale is based on the assumption that the hydroxyl and nitrate radicals as well as 2-NP are at Henry's equilibria in the gas phase and in the aqueous phase while the sulfate radical-anions exist only in the aqueous phase.

Table 3. Rate and Henry's constants for selected atmospheric reactants at 296–298 K.

Reaction X + NP	$k_{X,g}$		$k_{X,aq}$	H_{dX}	H_{dNP}
	cm^3 molecule^{-1} s^{-1}	dm^3 mol^{-1} s^{-1}	dm^3 mol^{-1} s^{-1}		
OH + 2-NP	9.00×10^{-13} (a)	5.42×10^8	5.90×10^9 (b)	6.11×10^2 (d)	5.17×10^3 (e)
NO_3 + 2-NP	2.00×10^{-14} (a)	1.20×10^7	2.30×10^7 (c)	1.47×10^1 (d)	

(a) [52,95]; (b) [54]; (c) [96]; (d) [81]; (e) [97].

Table 4. Ranges of radical concentrations in gas phase, clouds and deliquescent particles [94].

	Gas Phase [(a)]		Aqueous Phase				
	OH	NO_3	OH	NO_3	$SO_4^{\bullet-}$	$[OH]/[SO_4^{\bullet-}]$	$[NO_3]/[SO_4^{\bullet-}]$
	molecule cm^{-3}		mol dm^{-3}				
Minimal	1.4×10^2	7.9×10^6	1.4×10^{-16}	1.6×10^{-16}	5×10^{-17}	1.6×10^{-4}	1.8×10^{-3}
Maximal	6.6×10^3	1.2×10^7	8.0×10^{-12}	3.0×10^{-13}	9×10^{-13}	1.6×10^5	6.0×10^3

[(a)] calculated using Henry's constants from Table 3.

The rate of aqueous-phase reaction of a nitrophenol NP with sulfate radical-anions can also be compared to the rates of gas-phase reactions with OH or NO_3 sinks alone, assuming the gas-phase and aqueous-phase reactants follow the Henry's Law

$$R_{X,g-aq} = \frac{r_{X,g}}{r_{SO_4,aq}\omega} = \frac{k_{X,g}[X]_g[NP]_g}{k_{SO4,aq}[NP]_{aq}\left[SO_4^{-}\right]_{aq}\omega} = \frac{k_{X,g}}{k_{SO_4,aq}H_{d,X}H_{d,NP}\omega}\cdot\frac{[X]_{aq}}{\left[SO_4^{-}\right]_{aq}}, \quad (11)$$

The results obtained for 2-NP (Figure 6) show that the gas-phase sinks dominate in the aerosol and haze waters and in clouds and rains provided hydroxyl or nitrate radicals are in excess. Surprisingly, there is little difference observed between the hydroxyl radicals and nitrate radicals in spite of significant difference between the rate constants for their gas-phase reactions with 2-NP (Table 2). This is explained by lower solubility of NO_3 in water. When the aqueous-phase concentrations of both radicals are equal, the gas-phase concentration of NO_3 is higher than the concentration of OH and compensates for its lower rate constant. For reference, one can compare the gas-phase and the aqueous phase conversions of 2-NP by hydroxyl radicals or nitrite radicals alone (Figure S3). Conversion by OH in the gas phase dominates in over conversion in aerosol and haze water but not over that in clouds and rain. Conversion of 2-NP by NO_3 in the gas phase always dominates over conversion in atmospheric waters.

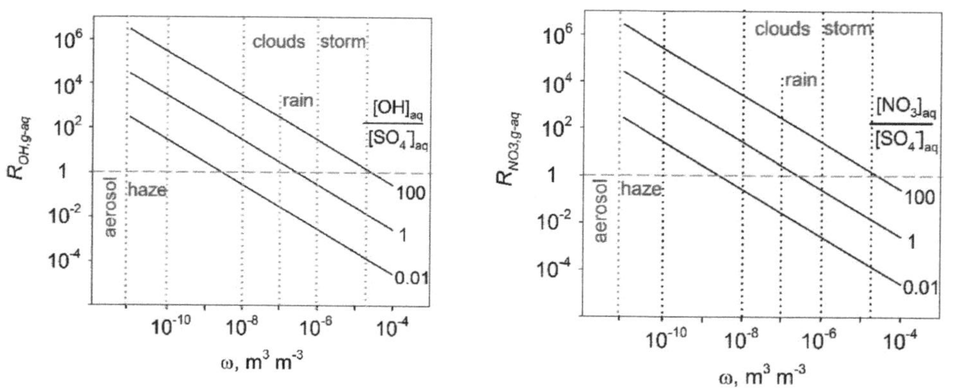

(a) (b)

Figure 6. Rate of 2-NP conversion in the atmosphere due to the gas-phase reaction with OH radicals (**a**) or with NO_3 radicals (**b**) compared to the rate of the aqueous-phase reaction of 2-NP with sulfate radical-anions for various proportions of radicals in the aqueous phase $[OH]_{aq}/[SO_4^{\bullet-}]_{aq}$ and varying liquid water contents (ω) (based on Equation (11)).

Rate of conversions of a NP by X (OH or NO_3) and by $SO_4^{\bullet-}$ in the aqueous phases alone were compared using Equation (12).

$$\frac{r_{X,aq}}{r_{SO_4,aq}} = \frac{k_{X,aq}}{k_{SO_4,aq}}\cdot\frac{[X]_{aq}}{\left[SO_4^{-}\right]_{aq}}, \quad (12)$$

Figure 7 shows that the aqueous-phase conversion of 2-NP by OH radicals dominates over that by $SO_4^{\bullet-}$ radical-anions when the ratio of the radicals concentrations is higher than ~0.15. A similar domination of NO_3 radicals over $SO_4^{\bullet-}$ radical-anions requires the corresponding concentration ratio is greater than ~40.

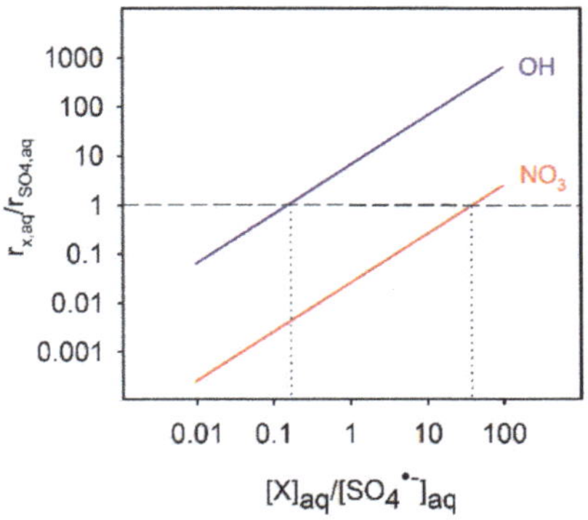

Figure 7. Comparison of the rates of aqueous-phase conversion of 2-NP due to reaction with $SO_4^{\bullet-}$ radicals and OH radicals (blue line) or NO_3 radicals (red line).

We expect the comparison of gas-phase and aqueous-phase conversions for other nitrophenols studied would provide similar results when the gas-phase rate constants for these nitrophenols are available.

5. Conclusions

Nitrophenols (2-NP, 3-NP, 4-NP, and 2,4-NP) react fast with $SO_4^{\bullet-}$ radical-anions in aqueous solutions. Rate constants for these reactions, along with rate constants of several chlorophenols and phenol, correlate linearly with Brown substituent coefficients and with the relative strength of the O–H bonds in the molecules. The correlation allows estimation of rate constants for reactions of other substituted phenols with sulfate radical-anions.

The aqueous-phase reaction of 2-NP with sulfate radical-anions dominates over the aqueous-phase conversion of 2-NP by OH radicals only when $SO_4^{\bullet-}$ radicals are at least 10 times more abundant than the OH radicals. Similar domination over NO_3 radical requires the concentration of sulfate radicals is at least a quarter of the concentration of nitrate radicals.

The comparison of gas-phase conversion of 2-NP by OH or NO_3 radicals against the aqueous-phase conversion by sulfate radical-anions depends on the liquid water contents of a particular atmospheric system considered. In deliquescent aerosol and haze water ($\omega < 10^{-10}$ m^3 m^{-3}), gas-phase reactions always prevail over the aqueous-phase reactions. In cloud, rain and fog water ($10^{-8} < \omega < 10^{-6}$ m^3 m^{-3}), the aqueous-phase reaction of 2-NP dominates over the gas-phase conversion of 2-NP by hydroxyl or nitrate radicals provided the aqueous-phase concentration of sulfate radical-anions is not smaller than the aqueous-phase concentration of hydroxyl or nitrate radicals. These conclusions are based on the assumption that the gas-phase and aqueous-phase concentrations of OH, NO_3, and 2-NP are bound by Henry's equilibria.

The gas-phase and aqueous-phase conversions of other nitrophenols are expected to follow similar patterns. However, this expectation should be confirmed by calculations when constants of the gas-phase reactions of the nitrophenols with hydroxyl and nitrate radicals are available.

Last not least, we hope that the rate constants determined in the present work for atmospheric purposes may appear useful for designers of advanced oxidation processes aimed at removal of nitrophenols from various waste effluents utilizing sulfate radical-anions.

Supplementary Materials: The following are available online at http://www.mdpi.com/2073-4433/10/12/795/s1, Table S1: Chain mechanism of SIV autoxidation catalyzed by FeIII, Figure S1: Concentration of oxygen recorded during autoxidation of $NaHSO_3$ solution in the presence of nitrophenols, Figure S2: Linear plots of reciprocal quasi-stationary rates for autoxidation of S(IV) in the presence of nitrophenols, Table S2: Properties of nitrophenols and sulfate radical anion, Figure S3: The ratio of the gas-phase and the aqueous-phase conversions of 2-NP by OH and NO_3 radicals.

Author Contributions: Conceptualization, methodology, investigation: K.J.R.; Data curation: K.J.R. and R.S.; Writing—original draft preparation: K.J.R.; Writing—review and editing: K.J.R. and R.S.

Funding: This research received no external funding.

Acknowledgments: The authors are grateful to Irena Grgić for the invitation to contribute to the Special Issue "Atmospheric Aqueous-Phase Chemistry" and to Józef Ziajka for extensive support in the experimental work behind this presentation.

Conflicts of Interest: The authors declare no conflict of interest.

References

1. Harrison, M.A.J.; Barra, S.; Borghesi, D.; Vione, D.; Arsene, C.; Iulian Olariu, R. Nitrated phenols in the atmosphere: A review. *Atmos. Environ.* **2005**, *39*, 231–248. [CrossRef]
2. Michałowicz, J.; Duda, W. Phenols—Sources and Toxicity. *Pol. J. Environ. Stud.* **2007**, *16*, 347–362.
3. Belloli, R.; Bolzacchini, E.; Clerici, L.; Rindone, B.; Sesana, G.; Librando, V. Nitrophenols in air and rainwater. *Environ. Eng. Sci.* **2006**, *23*, 405–415. [CrossRef]
4. Delhomme, O.; Morville, S.; Millet, M. Seasonal and diurnal variations of atmospheric concentrations of phenols and nitrophenols measured in the Strasbourg area, France. *Atmos. Pollut. Res.* **2010**, *1*, 16–22. [CrossRef]
5. Morville, S.; Scheyer, A.; Mirabel, P.; Millet, M. A multiresidue method for the analysis of phenols and nitrophenols in the atmosphere. *J. Environ. Monit.* **2004**, *6*, 963–966. [CrossRef]
6. Morville, S.; Scheyer, A.; Mirabel, P.; Millet, M. Spatial and geographical variations of urban, suburban and rural atmospheric concentrations of phenols and nitrophenols. *Environ. Sci. Pollut. Res.* **2006**, *13*, 83–89. [CrossRef]
7. Huston, R.; Chan, Y.C.; Gardner, T.; Shaw, G.; Chapman, H. Characterisation of atmospheric deposition as a source of contaminants in urban rainwater tanks. *Water Res.* **2009**, *43*, 1630–1640. [CrossRef]
8. Jaber, F.; Schummer, C.; Al Chami, J.; Mirabel, P.; Millet, M. Solid-phase microextraction and gas chromatography—Mass spectrometry for analysis of phenols and nitrophenols in rainwater, as their t-butyldimethylsilyl derivatives. *Anal. Bioanal. Chem.* **2007**, *387*, 2527–2535. [CrossRef]
9. Asman, W.A.H.; Jørgensen, A.; Bossi, R.; Vejrup, K.V.; Bügel Mogensen, B.; Glasius, M. Wet deposition of pesticides and nitrophenols at two sites in Denmark: Measurements and contributions from regional sources. *Chemosphere* **2005**, *59*, 1023–1031. [CrossRef]
10. Hofmann, D.; Hartmann, F.; Herrmann, H. Analysis of nitrophenols in cloud water with a miniaturized light-phase rotary perforator and HPLC-MS. *Anal. Bioanal. Chem.* **2008**, *391*, 161–169. [CrossRef]
11. Boris, A.J.; Lee, T.; Park, T.; Choi, J.; Seo, S.J.; Collett, J.L., Jr. Fog composition at Baengnyeong Island in the eastern Yellow Sea: Detecting markers of aqueous atmospheric oxidations. *Atmos. Chem. Phys.* **2016**, *16*, 437–453. [CrossRef]
12. Chow, K.S.; Huang, X.H.H.; Yu, J.Z. Quantification of nitroaromatic compounds in atmospheric fine particulate matter in Hong Kong over 3 years: Field measurement evidence for secondary formation derived from biomass burning emissions. *Environ. Chem.* **2016**, *13*, 665–673. [CrossRef]
13. Duong, H.T.; Kadokami, K.; Trinh, H.T.; Phan, T.Q.; Le, G.T.; Nguyen, D.T.; Nguyen, T.T.; Nguyen, D.T. Target screening analysis of 970 semi-volatile organic compounds adsorbed on atmospheric particulate matter in Hanoi, Vietnam. *Chemosphere* **2019**, *219*, 784–795. [CrossRef] [PubMed]
14. Fleming, L.T.; Lin, P.; Laskin, A.; Laskin, J.; Weltman, R.; Edwards, R.D.; Arora, N.K.; Yadav, A.; Meinardi, S.; Blake, D.R.; et al. Molecular composition of particulate matter emissions from dung and brushwood burning household cookstoves in Haryana, India. *Atmos. Chem. Phys.* **2018**, *18*, 2461–2480. [CrossRef]
15. Giorio, C.; Bortolini, C.; Kourtchev, I.; Tapparo, A.; Bogialli, S.; Kalberer, M. Direct target and non-target analysis of urban aerosol sample extracts using atmospheric pressure photoionisation high-resolution mass spectrometry. *Chemosphere* **2019**, *224*, 786–795. [CrossRef] [PubMed]

16. Irei, S.; Stupak, J.; Gong, X.; Chan, T.-W.; Cox, M.; McLaren, R.; Rudolph, J. Molecular marker study of particulate organic matter in Southern Ontario air. *J. Anal. Methods Chem.* **2017**, *2017*, 19. [CrossRef] [PubMed]
17. Kitanovski, Z.; Grgic, I.; Vermeylen, R.; Claeys, M.; Maenhaut, W. Liquid chromatography tandem mass spectrometry method for characterization of monoaromatic nitro-compounds in atmospheric particulate matter. *J. Chromatogr. A* **2012**, *1268*, 35–43. [CrossRef]
18. Kitanovski, Z.; Shahpoury, P.; Samara, C.; Voliotis, A.; Lammel, G. Composition and mass size distribution of nitrated and oxygenated aromatic compounds in ambient particulate matter from southern and central Europe—implications for origin. *Atmos. Chem. Phys. Discuss.* **2019**, *2019*, 1–29. [CrossRef]
19. Krivácsy, Z.; Gelencsér, A.; Kiss, G.; Mészáros, E.; Molnár, Á.; Hoffer, A.; Mészáros, T.; Sárvári, Z.; Temesi, D.; Varga, B.; et al. Study on the chemical character of water soluble organic compounds in fine atmospheric aerosol at the Jungfraujoch. *J. Atmos. Chem.* **2001**, *39*, 235–259. [CrossRef]
20. Li, X.; Jiang, L.; Hoa, L.P.; Lyu, Y.; Xu, T.; Yang, X.; Iinuma, Y.; Chen, J.; Herrmann, H. Size distribution of particle-phase sugar and nitrophenol tracers during severe urban haze episodes in Shanghai. *Atmos. Environ.* **2016**, *145*, 115–127. [CrossRef]
21. OÖzel, M.Z.; Hamilton, J.F.; Lewis, A.C. New sensitive and quantitative analysis method for organic nitrogen compounds in urban aerosol samples. *Environ. Sci. Technol.* **2011**, *45*, 1497–1505. [CrossRef] [PubMed]
22. Teich, M.; Pinxteren, D.; Herrmann, H. Determination of nitrophenolic compounds from atmospheric particles using hollow-fiber liquid-phase microextraction and capillary electrophoresis/mass spectrometry analysis. *Electrophoresis* **2014**, *35*, 1353–1361. [CrossRef] [PubMed]
23. Wang, L.; Wang, X.; Gu, R.; Wang, H.; Yao, L.; Wen, L.; Zhu, F.; Wang, W.; Xue, L.; Yang, L.; et al. Observations of fine particulate nitrated phenols in four sites in northern China: Concentrations, source apportionment, and secondary formation. *Atmos. Chem. Phys.* **2018**, *18*, 4349–4359. [CrossRef]
24. Williams, B.J.; Goldstein, A.H.; Kreisberg, N.M.; Hering, S.V.; Worsnop, D.R.; Ulbrich, I.M.; Docherty, K.S.; Jimenez, J.L. Major components of atmospheric organic aerosol in southern California as determined by hourly measurements of source marker compounds. *Atmos. Chem. Phys.* **2010**, *10*, 11577–11603. [CrossRef]
25. Zhang, Y.Y.; Müller, L.; Winterhalter, R.; Moortgat, G.K.; Hoffmann, T.; Pöschl, U. Seasonal cycle and temperature dependence of pinene oxidation products, dicarboxylic acids and nitrophenols in fine and coarse air particulate matter. *Atmos. Chem. Phys.* **2010**, *10*, 7859–7873. [CrossRef]
26. Kristanti, R.A.; Kanbe, M.; Toyama, T.; Tanaka, Y.; Tang, Y.; Wu, X.; Mori, K. Accelerated biodegradation of nitrophenols in the rhizosphere of Spirodela polyrrhiza. *J. Environ. Sci.* **2012**, *24*, 800–807. [CrossRef]
27. Vozňáková, Z.; Podehradská, J.; Kohličková, M. Determination of nitrophenols in soil. *Chemosphere* **1996**, *33*, 285–291. [CrossRef]
28. Chiron, S.; Minero, C.; Vione, D. Occurrence of 2,4-dichlorophenol and of 2,4-dichloro-6-nitrophenol in the Rhone River Delta (Southern France). *Environ. Sci. Technol.* **2007**, *41*, 3127–3133. [CrossRef]
29. Cho, E.; Khim, J.; Chung, S.; Seo, D.; Son, Y. Occurrence of micropollutants in four major rivers in Korea. *Sci. Total Environ.* **2014**, *491–492*, 138–147. [CrossRef]
30. Musilová, J.; Barek, J.; Pecková, K. Determination of nitrophenols in drinking and river water by differential pulse voltammetry at boron-doped diamond film electrode. *Electroanalysis* **2011**, *23*, 1236–1244. [CrossRef]
31. Schmidt-Bäumler, K.; Heberer, T.; Stan, H.-J. Occurrence and distribution of organic contaminants in the aquatic system in Berlin.Part II: Substituted phenols in Berlin surface water. *Acta Hydroch. Hydrob.* **1999**, *27*, 143–149. [CrossRef]
32. Balasubramanian, P.; Balamurugan, T.S.T.; Chen, S.-M.; Chen, T.-W. Simplistic synthesis of ultrafine CoMnO3 nanosheets: An excellent electrocatalyst for highly sensitive detection of toxic 4-nitrophenol in environmental water samples. *J. Hazard. Mater.* **2018**. [CrossRef]
33. Jay, K.; Stieglitz, L. Identification and quantification of volatile organic components in emissions of waste incineration plants. *Chemosphere* **1995**, *30*, 1249–1260. [CrossRef]
34. Goi, A.; Trapido, M. Comparison of advanced oxidation processes for the destruction of 2,4-dinitrophenol. *Proc. Est. Acad. Sci. Chem.* **2001**, *50*, 5–17.
35. Inomata, S.; Fushimi, A.; Sato, K.; Fujitani, Y.; Yamada, H. 4-Nitrophenol, 1-nitropyrene, and 9-nitroanthracene emissions in exhaust particles from diesel vehicles with different exhaust gas treatments. *Atmos. Environ.* **2015**, *110*, 93–102. [CrossRef]

36. Lu, C.; Wang, X.; Dong, S.; Zhang, J.; Li, J.; Zhao, Y.; Liang, Y.; Xue, L.; Xie, H.; Zhang, Q.; et al. Emissions of fine particulate nitrated phenols from various on-road vehicles in China. *Environ. Res.* **2019**, *179*, 108709. [CrossRef] [PubMed]
37. Rubio, M.A.; Bustamante, P.; Vásquez, P. Atmospheric phenolic derivatives as tracers in an urban area. *J. Chil. Chem. Soc.* **2019**, *64*, 4407–4411. [CrossRef]
38. Environmental Protection Agency. *Interim Reregistration Eligibility Decision for Methyl Parathion, Case No. 0153*; Office of Prevention, Pesticides and Toxic Substances, Ed.; United States Environmental Protection Agency: Washington, DC, USA, 2006; p. 20460.
39. World Health Organization. *Parathion in Drinking-Water*; Background document for development of WHO guidelines for drinking-water quality; World Health Organization: Geneva, Switzerland, 2004.
40. Bluvshtein, N.; Lin, P.; Flores, J.M.; Segev, L.; Mazar, Y.; Tas, E.; Snider, G.; Weagle, C.; Brown, S.S.; Laskin, A.; et al. Broadband optical properties of biomass-burning aerosol and identification of brown carbon chromophores. *J. Geophys. Res. Atmos.* **2017**, *122*, 5441–5456. [CrossRef]
41. Priestley, M.; Le Breton, M.; Bannan, T.J.; Leather, K.E.; Bacak, A.; Reyes-Villegas, E.; De Vocht, F.; Shallcross, B.M.A.; Brazier, T.; Anwar Khan, M.; et al. Observations of Isocyanate, Amide, Nitrate, and Nitro Compounds From an Anthropogenic Biomass Burning Event Using a ToF-CIMS. *J. Geophys. Res. Atmos.* **2018**, *123*, 7687–7704. [CrossRef]
42. Xie, M.; Chen, X.; Hays, M.D.; Holder, A.L. Composition and light absorption of N-containing aromatic compounds in organic aerosols from laboratory biomass burning. *Atmos. Chem. Phys.* **2019**, *19*, 2899–2915. [CrossRef]
43. Wang, X.; Gu, R.; Wang, L.; Xu, W.; Zhang, Y.; Chen, B.; Li, W.; Xue, L.; Chen, J.; Wang, W. Emissions of fine particulate nitrated phenols from the burning of five common types of biomass. *Environ. Pollut.* **2017**, *230*, 405–412. [CrossRef] [PubMed]
44. Hinrichs, R.Z.; Buczek, P.; Trivedi, J.J. Solar Absorption by Aerosol-Bound Nitrophenols Compared to Aqueous and Gaseous Nitrophenols. *Environ. Sci.Technol.* **2016**, *50*, 5661–5667. [CrossRef] [PubMed]
45. Lin, P.; Bluvshtein, N.; Rudich, Y.; Nizkorodov, S.A.; Laskin, J.; Laskin, A. Molecular Chemistry of Atmospheric Brown Carbon Inferred from a Nationwide Biomass Burning Event. *Environ. Sci.Technol.* **2017**, *51*, 11561–11570. [CrossRef] [PubMed]
46. Barzaghi, P.; Herrmann, H. A mechanistic study of the oxidation of phenol by $OH/NO_2/NO_3$ in aqueous solution. *Phys. Chem. Chem. Phys.* **2002**, *4*, 3669–3675. [CrossRef]
47. Heal, M.R.; Harrison, M.A.J.; Neil Cape, J. Aqueous-phase nitration of phenol by N_2O_5 and $ClNO_2$. *Atmos. Environ.* **2007**, *41*, 3515–3520. [CrossRef]
48. Vione, D.; Maurino, V.; Minero, C.; Pelizzetti, E. Aqueous atmospheric chemistry: Formation of 2,4-dinitrophenol upon nitration of 2-nitrophenol and 4-nitrophenol in solution. *Environ. Sci. Technol.* **2005**, *39*, 7921–7931. [CrossRef]
49. Vione, D.; Maurino, V.; Minero, C.; Pelizzetti, E.; Harrison, M.A.J.; Olariu, R.; Arsene, C. Photochemical reactions in the tropospheric aqueous phase and on particulate matter. *Chem. Soc. Rev.* **2006**, *35*, 441–453. [CrossRef]
50. Yuan, B.; Liggio, J.; Wentzell, J.; Li, S.M.; Stark, H.; Roberts, J.M.; Gilman, J.; Lerner, B.; Warneke, C.; Li, R.; et al. Secondary formation of nitrated phenols: Insights from observations during the Uintah Basin Winter Ozone Study (UBWOS) 2014. *Atmos. Chem. Phys.* **2016**, *16*, 2139–2153. [CrossRef]
51. Atkinson, R. Gas-phase tropospheric chemistry of organic compounds. *J. Phys. Chem. Ref. Data Monogr.* **1994**, Monoghraph No. 2, 1–216.
52. Atkinson, R.; Aschmann, S.M.; Arey, J. Reactions of hydroxyl and nitrogen trioxide radicals with phenol, cresols, and 2-nitrophenol at 296 ± 2 K. *Environ. Sci.Technol.* **1992**, *26*, 1397–1403. [CrossRef]
53. Barsotti, F.; Bartels-Rausch, T.; De Laurentiis, E.; Ammann, M.; Brigante, M.; Mailhot, G.; Maurino, V.; Minero, C.; Vione, D. Photochemical Formation of Nitrite and Nitrous Acid (HONO) upon Irradiation of Nitrophenols in Aqueous Solution and in Viscous Secondary Organic Aerosol Proxy. *Environ. Sci.Technol.* **2017**, *51*, 7486–7495. [CrossRef] [PubMed]
54. Vione, D.; Maurino, V.; Minero, C.; Duncianu, M.; Olariu, R.-I.; Arsene, C.; Sarakha, M.; Mailhot, G. Assessing the transformation kinetics of 2-and 4-nitrophenol in the atmospheric aqueous phase. Implications for the distribution of both nitroisomers in the atmosphere. *Atmos. Environ.* **2009**, *43*, 2321–2327. [CrossRef]

55. Hems, R.F.; Abbatt, J.P.D. Aqueous Phase Photo-oxidation of Brown Carbon Nitrophenols: Reaction Kinetics, Mechanism, and Evolution of Light Absorption. *ACS Earth Space Chem.* **2018**, *2*, 225–234. [CrossRef]
56. Li, Y.J.; Huang, D.D.; Cheung, H.Y.; Lee, A.K.Y.; Chan, C.K. Aqueous-phase photochemical oxidation and direct photolysis of vanillin—A model compound of methoxy phenols from biomass burning. *Atmos. Chem. Phys.* **2014**, *14*, 2871–2885. [CrossRef]
57. Wei, B.; Sun, J.; Mei, Q.; He, M. Mechanism and kinetic of nitrate radical-initiated atmospheric reactions of guaiacol (2-methoxyphenol). *Comp. Theor. Chem.* **2018**, *1129*, 1–8. [CrossRef]
58. Yu, L.; Smith, J.; Laskin, A.; Anastasio, C.; Laskin, J.; Zhang, Q. Chemical characterization of SOA formed from aqueous-phase reactions of phenols with the triplet excited state of carbonyl and hydroxyl radical. *Atmos. Chem. Phys.* **2014**, *14*, 13801–13816. [CrossRef]
59. Yu, L.; Smith, J.; Laskin, A.; George, K.M.; Anastasio, C.; Laskin, J.; Dillner, A.M.; Zhang, Q. Molecular transformations of phenolic SOA during photochemical aging in the aqueous phase: Competition among oligomerization, functionalization, and fragmentation. *Atmos. Chem. Phys.* **2016**, *16*, 4511–4527. [CrossRef]
60. Litter, M.I. Introduction to Photochemical Advanced Oxidation Processes for Water Treatment. In *Environmental Photochemistry Part II*; Boule, P., Bahnemann, D.W., Robertson, P.K.J., Eds.; Springer: Berlin/Heidelberg, Germany, 2005; Volume 2, Pt M, pp. 325–366.
61. Goi, A.; Trapido, M. Hydrogen peroxide photolysis, Fenton reagent and photo-Fenton for the degradation of nitrophenols: A comparative study. *Chemosphere* **2002**, *46*, 913–922. [CrossRef]
62. Ahn, W.-Y.; Sheeley, S.A.; Rajh, T.; Cropek, D.M. Photocatalytic reduction of 4-nitrophenol with arginine-modified titanium dioxide nanoparticles. *Appl. Catal. B Environ.* **2007**, *74*, 103–110. [CrossRef]
63. Ali, I.; Hassan, A.; Shabaan, S.; El-Nasser, K. Synthesis and characterization of composite catalysts Cr/ZSM-5 and their effects toward photocatalytic degradation of p-nitrophenol. *Arab. J. Chem.* **2017**, *10*, S2106–S2114. [CrossRef]
64. Priya, M.H.; Madras, G. Kinetics of photocatalytic degradation of chlorophenol, nitrophenol, and their mixtures. *Ind. Eng. Chem. Res.* **2006**, *45*, 482–486. [CrossRef]
65. Liu, Y.; Liu, H.; Li, Y. Comparative study of the electrocatalytic oxidation and mechanism of nitrophenols at Bi-doped lead dioxide anodes. *Appl. Catal. B Environ.* **2008**, *84*, 297–302. [CrossRef]
66. Christensen, P.A.; Egerton, T.A.; Kosa, S.A.M.; Tinlin, J.R.; Scott, K. The photoelectrocatalytic oxidation of aqueous nitrophenol using a novel reactor. *J. Appl. Electrochem.* **2005**, *35*, 683–692. [CrossRef]
67. Zhou, M.; Lei, L. An improved UV/Fe^{3+} process by combination with electrocatalysis for p-nitrophenol degradation. *Chemosphere* **2006**, *63*, 1032–1040. [CrossRef] [PubMed]
68. Apolinário, Â.C.; Silva, A.M.T.; Machado, B.F.; Gomes, H.T.; Araújo, P.P.; Figueiredo, J.L.; Faria, J.L. Wet air oxidation of nitro-aromatic compounds: Reactivity on single-and multi-component systems and surface chemistry studies with a carbon xerogel. *Appl. Catal. B Environ.* **2008**, *84*, 75–86. [CrossRef]
69. Diaz, E.; Polo, A.M.; Mohedano, A.F.; Casas, J.A.; Rodriguez, J.J. On the biodegradability of nitrophenols and their reaction products by catalytic hydrogenation. *J. Chem. Technol. Biotechnol.* **2012**, *87*, 1263–1269. [CrossRef]
70. Anipsitakis, G.P.; Dionysiou, D.D.; Gonzalez, M.A. Cobalt-Mediated Activation of Peroxymonosulfate and Sulfate Radical Attack on Phenolic Compounds. Implications of Chloride Ions. *Environ. Sci. Technol.* **2006**, *40*, 1000–1007. [CrossRef]
71. EPA. The Original List of Hazardous Air Pollutants. Available online: http://www.epa.gov/ttn/atw/188polls.html (accessed on 29 October 2014).
72. EPA. Priority Pollutants. Available online: http://water.epa.gov/scitech/methods/cwa/pollutants.cfm (accessed on 27 October 2014).
73. Keith, L.; Telliard, W. ES&T Special Report: Priority pollutants: I-a perspective view. *Environ. Sci.Technol.* **1979**, *13*, 416–423. [CrossRef]
74. Gramatica, P.; Santagostino, A.; Bolzacchini, E.; Rindone, B. Atmospheric monitoring, toxicology and QSAR modelling of nitrophenols. *Fresenius Environ. Bull.* **2002**, *11*, 757–762.
75. Schafer, K.S.; Reeves, M.; Spitzer, S.; Kegley, S.E. *Chemical Trespass Pesticides in our Bodies and Corporate Accountability*; Pesticide Action Network North America: San Francisco, CA, USA, 2004.
76. Zhang, D.-P.; Wu, W.-L.; Long, H.-Y.; Liu, Y.-C.; Yang, Z.-S. Voltammetric Behavior of o-Nitrophenol and Damage to DNA. *Int. J. Mol. Sci.* **2008**, *9*, 316–326. [CrossRef]

77. Buxton, G.V.; Salmon, G.A.; Williams, J.E. The reactivity of biogenic monoterpenes towards OH·and SO_4^- radicals in de-oxygenated acidic solution. *J. Atmos. Chem.* **2000**, *36*, 111–134. [CrossRef]
78. Clifton, C.L.; Huie, R.E. Rate constants for hydrogen abstraction reactions of the sulfate radical, SO_4^-. Alcohols. *Int. J. Chem. Kinet.* **1989**, *21*, 677–687. [CrossRef]
79. George, C.; Rassy, H.E.; Chovelon, J.M. Reactivity of selected volatile organic compounds (VOCs) toward the sulfate radical (SO_4^-). *Int. J. Chem. Kinet.* **2001**, *33*, 539–547. [CrossRef]
80. Herrmann, H.; Hoffmann, D.; Schaefer, T.; Bräuer, P.; Tilgner, A. Tropospheric Aqueous-Phase Free-Radical Chemistry: Radical Sources, Spectra, Reaction Kinetics and Prediction Tools. *ChemPhysChem* **2010**, *11*, 3796–3822. [CrossRef] [PubMed]
81. Rudzinski, K.J. Degradation of Isoprene in the Presence of Sulphoxy Radical Anions. *J. Atmos. Chem.* **2004**, *48*, 191–216. [CrossRef]
82. Rudziński, K.J.; Gmachowski, L.; Kuznietsova, I. Reactions of isoprene and sulphoxy radical-anions—A possible source of atmospheric organosulphites and organosulphates. *Atmos. Chem. Phys.* **2009**, *9*, 2129–2140. [CrossRef]
83. Ziajka, J.; Rudzinski, K. Autoxidation of S-IV inhibited by chlorophenols reacting with sulfate radicals. *Environ. Chem.* **2007**, *4*, 355–363. [CrossRef]
84. Ziajka, J.; Pasiuk-Bronikowska, W. Autoxidation of sulphur dioxide in the presence of alcohols under conditions related to the tropospheric aqueous phase. *Atmos. Environ.* **2003**, *37*, 3913–3922. [CrossRef]
85. Ziajka, J.; Pasiuk-Bronikowska, W. Rate constants for atmospheric trace organics scavenging SO_4^- in the Fe-catalysed autoxidation of S(IV). *Atmos. Environ.* **2005**, *39*, 1431–1438. [CrossRef]
86. Grgić, I.; Podkrajšek, B.; Barzaghi, P.; Herrmann, H. Scavenging of SO_4^- radical anions by mono-and dicarboxylic acids in the Mn(II)-catalyzed S(IV) oxidation in aqueous solution. *Atmos. Environ.* **2007**, *41*, 9187–9194. [CrossRef]
87. Elliot, A.J.; McCracken, D.R.; Buxton, G.V.; Wood, N.D. Estimation of rate constants for near-diffusion-controlled reactions in water at high temperatures. *J. Chem. Soc. Faraday Trans.* **1990**, *86*, 1539–1547. [CrossRef]
88. Zhu, L.; Nicovich, J.M.; Wine, P.H. Temperature-dependent kinetics studies of aqueous phase reactions of SO_4^- radicals with dimethylsulfoxide, dimethylsulfone, and methanesulfonate. *J. Photochem. Photobiol. A* **2003**, *157*, 311–319. [CrossRef]
89. Noyes, R.M. Effects of diffusion rates on chemical kinetics. In *Progress in Reaction Kinetics;* Porter, G., Stevens, B., Eds.; Pergamon Press: Oxford, UK, 1961; Volume 1, pp. 129–160.
90. North, A.M. *The Collision Theory of Chemical Reactions in Liquids;* Methuen: London, UK, 1964.
91. Hansch, C.; Leo, A.; Taft, R.W. A survey of Hammett substituent constants and resonance and field parameters. *Chem. Rev.* **1991**, *91*, 165–195. [CrossRef]
92. Jones, R.A.Y. *Physical and Mechanistic Organic Chemistry;* Cambridge University Press: Cambridge, MA, USA, 1979.
93. Jonsson, M.; Lind, J.; Eriksen, T.E.; Merényi, G. O–H bond strengths and one-electron reduction potentials of multisubstituted phenols and phenoxyl radicals. Predictions using free energy relationships. *J. Chem. Soc. Perkin Trans. 2* **1993**, 1567–1568. [CrossRef]
94. Herrmann, H.; Schaefer, T.; Tilgner, A.; Styler, S.A.; Weller, C.; Teich, M.; Otto, T. Tropospheric aqueous-phase chemistry: Kinetics, mechanisms, and its coupling to a changing gas phase. *Chem. Rev.* **2015**, *115*, 4259–4334. [CrossRef]
95. Atkinson, R. Kinetics and mechanisms of the gas-phase reactions of the hydroxyl radical with organic compounds. *Chem. Ref.* **1989**, Monograph No. 1. 1–246.
96. Herrmann, H. Kinetics of Aqueous Phase Reactions Relevant for Atmospheric Chemistry. *Chem. Rev.* **2003**, *103*, 4691–4716. [CrossRef]
97. Guo, X.X.; Brimblecombe, P. Henry's law constants of phenol and mononitrophenols in water and aqueous sulfuric acid. *Chemosphere* **2007**, *68*, 436–444. [CrossRef]

atmosphere

Article

pH- and Temperature-Dependent Kinetics of the Oxidation Reactions of OH with Succinic and Pimelic Acid in Aqueous Solution

Thomas Schaefer, Liang Wen, Arne Estelmann, Joely Maak and Hartmut Herrmann *

Atmospheric Chemistry Department (ACD), Leibniz-Institute for Tropospheric Research (TROPOS), Permoserstrasse 15, 04318 Leipzig, Germany

* Correspondence: herrmann@tropos.de; Tel.: +49-341-2717-7024

Received: 4 March 2020; Accepted: 24 March 2020; Published: 26 March 2020

Abstract: Rate constants for the aqueous-phase reactions of the hydroxyl radical with the dicarboxylic acids, succinic acid and pimelic acid were determined using the relative rate technique over the temperature range 287 K ≤ T ≤ 318 K and at pH = 2.0, 4.6 or 4.9 and 8.0. OH radicals were generated by H_2O_2 laser flash photolysis while thiocyanate was used as a competitor. The pH values were adjusted to obtain the different speciation of the dicarboxylic acids. The following Arrhenius expressions were determined (in units of L mol^{-1} s^{-1}): succinic acid, $k(T, AH_2)$ (2.1 ± 0.1) × 10^{10} exp[(−1530 ± 250 K)/T], $k(T, AH^-)$ (1.8 ± 0.1) × 10^{10} exp[(−1070 ± 370 K)/T], $k(T, A^{2-})$ (2.9 ± 0.2) × 10^{11} exp[(−1830 ± 350 K)/T] and pimelic acid, $k(T, AH_2)$ (7.3 ± 0.2) × 10^{10} exp[(−1040 ± 140 K)/T], $k(T, AH^-)$ (1.8 ± 0.1) × 10^{11} exp[(−1200 ± 240 K)/T], $k(T, A^{2-})$ (1.4 ± 0.1) × 10^{12} exp[(−1830 ± 110 K)/T]. A general OH radical reactivity trend for dicarboxylic acids was found as $k(AH_2) < k(AH^-) < k(A^{2-})$. By using the pH and temperature dependent rate constants, source and sinking processes in the tropospheric aqueous phase can be described precisely.

Keywords: OH radicals; rate constants; dicarboxylic acids; succinic acid; pimelic acid; aqueous phase chemistry

1. Introduction

Dicarboxylic acids such as oxalic, succinic, and pimelic acid are important trace constituents in the troposphere where they exist predominately in the particle phase as demonstrated by the Henry's Law constant Hcp = (2–4.1) × 10^7 mol m^{-3} Pa^{-1} for succinic acid and Hcp = 8.1 × 10^7 mol m^{-3} Pa^{-1} for pimelic acid [1,2]. Dicarboxylic acids are ubiquitous and have been found in remote and polluted areas as well as in pristine areas such as the Arctic and Antarctic region [3–8]. These compounds reach the pristine areas through long-range transport or originate from the photochemical oxidation of volatile organic compounds (VOCs) (e.g., isoprene), biogenic unsaturated fatty acids emitted from the surface ocean [4], or biomass burning or vehicle emissions [3,4]. Organic acids and diacids can contribute to the acidity of cloud water [9,10], lead to the secondary organic aerosol (SOA) formation by fostering accretion reactions such as hemiacetal formation or esterification [11]), and catalyze the oxidation of SO_2 [12] and the hydrolysis of SO_3 [13]. The oxidation of dicarboxylic acid generally lead either to more functionalized (substituted) acids or to the degradation by shortening the carbon chain [7,11,14–17]. The aqueous-phase oxidation kinetics of carboxylic acids and dicarboxylic acid towards hydroxyl radicals have been investigated in several studies in the past [14,17–24]. These investigations clearly show an influence of the speciation of the dicarboxylic acid on the OH radical rate constant, a systematic investigation of the monoanionic form of the carboxylic acid is still missing. This motivates further investigations of the temperature- and pH-dependent of hydroxyl radical reaction with succinic acid (C_4) and pimelic acid (C_7) studied here, since (i) these DCAs have been identified in ambient aerosol particles [4,25,26] and (ii) the pH in the tropospheric aqueous phases varies between −2.5 and +9.35 [9]. The results were obtained by using the laser flash photolysis of hydrogen peroxide to generate the OH radicals, which react with a competitor SCN^-. This relative rate technique is suitable over large temperature ranges and can be applied at different pH values. Dicarboxylic acids have two acid-base equilibria, so the pH value was chosen in such a way to first separately investigate the fully protonated and the fully deprotonated forms, and then determine the rate constant of the singly deprotonated form in order to characterize their atmospheric lifetime.

2. Experiments

2.1. Materials

The following chemicals were used as received: hydrogen peroxide (≥30% in water, Chemsolute), KSCN (≥99.0%, Chemsolute), Succinic acid (≥99.0%, Sigma-Aldrich), Pimelic acid (≥98.0%, Fluka), $HClO_4$ (70–72%, J. T. Baker Analyzed), NaOH (1 mol L^{-1} in water, Fluka). All solutions were freshly prepared using Milli-Q grade water (18 MΩ cm, Millipore, Billerica, MA).

2.2. Methods

The second order OH radical rate constants of the reaction with organic acids were determined by applying the thiocyanate competition kinetics method [27–29] using a temperature-controlled laser flash photolysis long path absorption (LP-LPA) setup [28,29].

2.2.1. Experimental Setup

Briefly, the setup includes an excimer laser (Compex 201, Lambda-Physik) operating with a 20 ns pulse and a frequency of f = 4 Hz at λ = 248 nm as photolysis light source, a cuboid reaction flow cell (28 mL volume) with a photolysis path length of 3.5 cm, and a White cell mirror configuration to increase the optical path length of the analytical light to d = 48 cm. Two continuous-wave lasers at λ = 407 nm (Radius 405, Coherent) or λ = 561 nm (LCX-561, Oxxius) were used as analyzing light sources. A photodiode (S1336-44BQ, Hamamatsu) was applied as detector. For spectrometric measurements, a double-beam spectrometer (Lambda 900, Perkin Elmer) was applied and the pH value of the measurement solution was measured with a pH-electrode (N 5900 A, SI Analytics).

2.2.2. Competition Kinetics Method

In a competition reaction of the OH radical between the organic and the reference reactant thiocyanate an absorbing radical $(SCN)_2^-$ is formed in reactions (R-3) to (R-5). The photolysis of hydrogen peroxide (R-1) ($c(H_2O_2) = 2 \cdot 10^{-4}$ mol L^{-1}) by a laser pulse at λ = 248 nm leads to the formation of OH radicals.

$$H_2O_2 + h\nu\ (248\ \text{nm}) \rightarrow 2\ OH \quad \text{(R-1)}$$

$$OH + R\text{-}H \rightarrow R + H_2O \quad \text{(R-2)}$$

$$OH + SCN^- \rightleftarrows SCNOH^- \quad \text{(R-3)}$$

$$SCNOH^- \rightleftarrows SCN + OH^- \quad \text{(R-4)}$$

$$SCN + SCN^- \rightleftarrows (SCN)_2^- \quad \text{(R-5)}$$

$$(SCN)_2^- + (SCN)_2^- \rightleftarrows (SCN)_2 + 2\ SCN^- \quad \text{(R-6)}$$

Further details of the thiocyanate reference reactant in the competition kinetics method can be found in Herrmann et al. [14] and the references therein. The OH concentration can be given in a range of $c(OH) = 1 \cdot 10^{-7}\text{–}1 \cdot 10^{-6}$ mol L^{-1}. The change of the absorption (cf. Figure S1 in the Supplementary Materials), due to the formation of $(SCN)_2^-$ radicals, was monitored at 407 or 561 nm ($\varepsilon_{407\ nm}$ = 3425 mol L^{-1} cm^{-1} and $\varepsilon_{561\ nm}$ = 3315 L mol^{-1} cm^{-1}) [14]. Each absorbance-time profile in Figure S1 was obtained by averaging 8 × 32 single traces. The absorbance error was determined by using the standard deviation and the Student's t-distribution. The OH radical reaction rate constants of the dicarboxylic acids (k_{RH}) can be derived, by using Equation (1).

$$\frac{A[(SCN)_2^-]_{0,X}}{A[(SCN)_2^-]_X} = 1 + \frac{k_{RH}[RH]}{k_0[SCN^-]} \quad (1)$$

The expressions $A[(SCN)_2^-]_{0,x}$ and $A[(SCN)_2^-]_x$ corresponds to the maximum absorption of the $(SCN)_2^-$ in the absence and in the presence of the organic reactant. The concentration of the organic reactant (c(RH) = $2.5 \cdot 10^{-5}$ –$4.0 \cdot 10^{-4}$ mol L^{-1}) is increased stepwise, while the reference reactant concentration (c(SCN^-) = $2.0 \cdot 10^{-5}$ mol L^{-1}) is constant in the series of five single measurement. Both concentrations, that of the organic reactant and that of the reference reactant, are at least 10 times higher than the concentration of the OH radicals in order to fulfil the conditions of the pseudo-first order. The ratios of $A[(SCN)_2^-]_{0,x}/A[(SCN)_2^-]_x$ were plotted against the corresponding ratios of [RH]/[SCN^-] (cf. Figure S2 in the Supplementary Materials) and from the resulting slope the second order rate constant can be calculated by using the temperature-dependent reference rate constants (k_0) by Zhu et al. [27]. The reported errors of the rate constant represent the statistical error for a 95% confidence interval.

3. Results and Discussion

In the present study, the pH- and T-dependencies of the second order rate constants of the reactions of OH with the organic diacids succinic (C_4 dicarboxylic acid) and pimelic acid (C_7 dicarboxylic acid) were determined. Table 1 contains the determined OH radical rate constants at three different pH values (pH = 2.0, 4.6 or 4.9, and 8) and five different temperatures (287 K ≤ T ≤ 318 K).

Table 1. Observed OH radical second order rate constants of the reaction with studied dicarboxylic acids at different temperatures.

Compounds and pH Values	$k_{(278\ K)}$/L mol^{-1} s^{-1}	$k_{(288\ K)}$/L mol^{-1} s^{-1}	$k_{(298\ K)}$/L mol^{-1} s^{-1}	$k_{(308\ K)}$/L mol^{-1} s^{-1}	$k_{(318\ K)}$/L mol^{-1} s^{-1}
			Succinic acid		
pH = 2	$(7.9 \pm 0.4) \cdot 10^7$	$(1.2 \pm 0.1) \cdot 10^8$	$(1.3 \pm 0.2) \cdot 10^8$	$(1.4 \pm 0.4) \cdot 10^8$	$(1.7 \pm 0.1) \cdot 10^8$
pH = 8⁻	$(4.3 \pm 0.1) \cdot 10^8$	$(4.8 \pm 0.1) \cdot 10^8$	$(5.3 \pm 0.2) \cdot 10^8$	$(8.5 \pm 0.3) \cdot 10^8$	$(9.2 \pm 0.1) \cdot 10^8$
pH = 4.9	$(2.9 \pm 0.1) \cdot 10^8$	$(3.6 \pm 0.1) \cdot 10^8$	$(4.6 \pm 0.3) \cdot 10^8$	$(4.5 \pm 0.3) \cdot 10^8$	$(3.6 \pm 0.1) \cdot 10^8$
			Pimelic acid		
pH = 2	$(1.7 \pm 0.1) \cdot 10^9$	$(2.2 \pm 0.1) \cdot 10^9$	$(2.2 \pm 0.1) \cdot 10^9$	$(2.5 \pm 0.1) \cdot 10^9$	$(2.8 \pm 0.1) \cdot 10^9$
pH = 8⁻	$(1.9 \pm 0.1) \cdot 10^9$	$(2.5 \pm 0.1) \cdot 10^9$	$(3.2 \pm 0.1) \cdot 10^9$	$(3.8 \pm 0.1) \cdot 10^9$	$(4.4 \pm 0.1) \cdot 10^9$
pH = 4.6[1]	$(2.1 \pm 0.1) \cdot 10^9$	$(2.3 \pm 0.1) \cdot 10^9$	$(2.8 \pm 0.1) \cdot 10^9$	$(3.2 \pm 0.1) \cdot 10^9$	$(3.5 \pm 0.1) \cdot 10^9$

The pK_a values of the diacids are available for succinic acid as $pK_{a,1}$ = 4.207 and $pK_{a,2}$ = 5.363 and for pimelic acid as $pK_{a,1}$ = 4.49 and $pK_{a,2}$ = 5.43 at ionic strength zero [30].

The ratio of the different forms AH_2, AH^-, and A^{2-} as a function of the pH value shown in Figure 1 (top) for succinic acid and in Figure 1 (bottom) for pimelic acid can be calculated. The maximum of the monoanionic dicarboxylic acid form AH^- is present at pH 4.9 for both the investigated organic diacids. Following the acid–base equilibria in Figure 1, the OH radical rate constant measurement of k(obs), was performed close to the pH of the maximum occurrence of the AH^- form. To derive the rate constants of the monoanionic form, k(AH^-), the observed rate constant k(obs) has to be corrected for the contribution χ of the fully protonated form AH_2 at pH = 2 as well as by the fully deprotonated form A^{2-} at pH = 8 using the following equation (Equation (2)). The term χ describes the percentage ratio of speciation calculated from the acid–base equilibria.

$$k(obs) = k(AH^-) \cdot \chi(AH^-) + k(A^{2-}) \cdot \chi(A^{2-}) + k(AH_2) \cdot \chi(AH_2) \ /L\,mol^{-1}s^{-1} \tag{2}$$

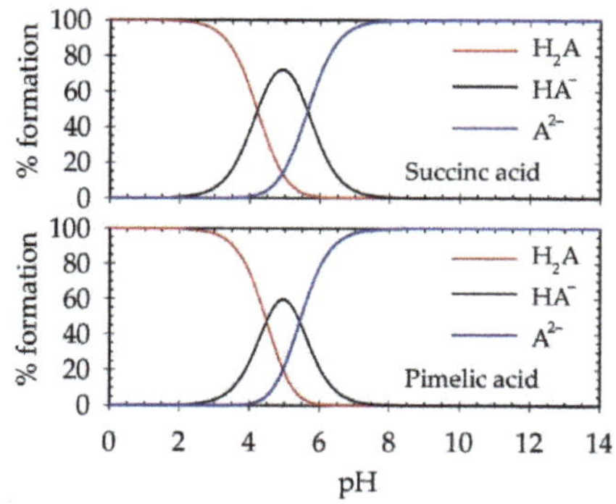

Figure 1. Speciation AH_2, AH^-, and A^{2-} of succinic acid (**top**) and of pimelic acid (**bottom**) as a function of the pH value, prepared with data of [30].

In addition, the T-dependency of the pK_a value treated by Equation (3), derived from the van't Hoff equation [28] (Equation (4)) shows only a small influence of the temperature on the pK_a value (Table 2).

$$pK_a = pK_a^\circ + \frac{\Delta_R H^\circ}{R \ln 10}\left(\frac{1}{T} - \frac{1}{T^\circ}\right) \tag{3}$$

$$\left(\frac{\partial \ln K}{\partial T}\right)_p = \frac{\Delta_R H^\circ}{RT^2} \tag{4}$$

Table 2. Calculated pK_a values and radical rate constants of the monoanionic dicarboxylic acids AH^- at different temperatures.

Compounds, pK_a Values and Rate Constants	278 K	288 K	298 K	308 K	318 K
			Succinic acid		
$pK_{a,1}$	4.243	4.224	4.207	4.191	4.176
$pK_{a,2}$	5.634	5.635	5.636	5.637	5.638
$k_{2nd}(AH^-)$/L mol^{-1} s^{-1}	$(3.0 \pm 0.1) \cdot 10^8$	$(3.9 \pm 0.1) \cdot 10^8$	$(5.1 \pm 0.1) \cdot 10^8$	$(4.4 \pm 0.1) \cdot 10^8$	$(5.3 \pm 0.1) \cdot 10^8$
			Pimelic acid		
$pK_{a,1}$	4.474	4.482	4.490	4.497	4.504
$pK_{a,2}$	5.383	5.407	5.430	5.451	5.471
$k_{2nd}(AH^-)$/L mol^{-1} s^{-1}	$(2.5 \pm 0.1) \cdot 10^9$	$(2.4 \pm 0.1) \cdot 10^9$	$(3.4 \pm 0.1) \cdot 10^9$	$(3.6 \pm 0.1) \cdot 10^9$	$(4.0 \pm 0.1) \cdot 10^9$

The standard enthalpy $\Delta_R H^\circ$ at T = 298.15 K for succinic acid can be given $\Delta_R H^\circ = -2845.12$ J mol^{-1} at $pK_{a,1} = 4.207$ and $\Delta_R H^\circ = 167.36$ J mol^{-1} at $pK_{a,2} = 5.636$ as well as for pimelic acid given with $\Delta_R H^\circ = 1255.2$ J mol^{-1} at $pK_{a,1} = 4.49$ and $\Delta_R H^\circ = 3765.6$ J mol^{-1} at $pK_{a,2} = 5.43$ [30]. In order to calculate the rate constant for the reaction of OH with the monoanion form as exactly as possible from the observed rate constant at pH = 4.9 and 4.6 the T-dependent pK_a values given in Table 2 were used.

The UV spectra of the dicarboxylic acids were measured to clarify a possible influence of the internal absorption effect [28] on the measured rate constants. Since the obtained molar attenuation coefficient at the excitation wavelength $\lambda = 248$ nm is lower than $\varepsilon_{248\,nm} < 1$ L mol^{-1} cm^{-1}, this effect can be neglected for the used measurement conditions.

Briefly, the diffusion limit rate constants of the OH radical reaction were calculated by using a method based on the Smoluchowski equation [31]. Further details of this method can be found elsewhere [32–34]. The obtained values (unit/L mol^{-1} s^{-1}) $k_{diff.}$(278 K) = 0.7 × 10^{10}, $k_{diff.}$(288 K) = 1.0 × 10^{10}, $k_{diff.}$(298 K) = 1.4 × 10^{10}, $k_{diff.}$(308 K) = 1.7 × 10^{10}, and $k_{diff.}$(318 K) = 2.2 × 10^{10} show that the measured rate constants are far away from the diffusion limitation and hence are chemically controlled.

The temperature-dependency of these radical rate constants can be described by an Arrhenius type equation [35] (Equation (5)), from this the pre-exponential factor, A, and the activation energy, E_A, are derived. The parameters R and T describe the gas constant and the temperature.

$$k(T) = A \cdot \exp(-E_A/RT)\ /L\ mol^{-1}\ s^{-1} \quad (5)$$

The T-dependent rate constants of the OH radical reaction with succinic acid were determined and are depicted in Figure 2.

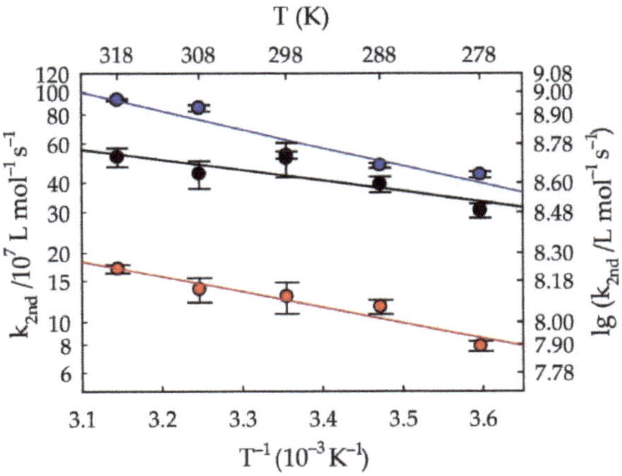

Figure 2. Arrhenius plot of the OH radical reactions of the speciation AH_2 (red), AH^- (black), and A^{2-} (blue) of succinic acid.

The derived Arrhenius expressions are shown in Equation (6) for the protonated succinic acid (pH = 2), in Equation (6) for the monoanionic form (pH = 4.9), and in Equation (8) for the dianionic form (pH = 8.0).

$$k(T, AH_2) = (2.1 \pm 0.1) \cdot 10^{10} \cdot \exp\left(\frac{-1530 \pm 250\ K}{T}\right) /L\ mol^{-1}\ s^{-1} \quad (6)$$

$$k(T, AH^-) = (1.8 \pm 0.1) \cdot 10^{10} \cdot \exp\left(\frac{-1070 \pm 370\ K}{T}\right) /L\ mol^{-1}\ s^{-1} \quad (7)$$

$$k(T, A^{2-}) = (2.9 \pm 0.2) \cdot 10^{11} \cdot \exp\left(\frac{-1830 \pm 350\ K}{T}\right) /L\ mol^{-1}\ s^{-1} \quad (8)$$

The temperature-dependency of the rate constants for the OH radical reaction with succinic acid can be described in the following order: AH_2 < AH^- < A^{2-}. Under alkaline conditions, the reaction is more T-dependent than under acidic conditions as it can be seen in the Arrhenius plot and the derived Arrhenius expressions.

The obtained Arrhenius plot of pimelic acid oxidation by OH radicals at pH = 2.0, pH = 4.6 and pH = 8 is shown in Figure 3. From the Arrhenius plot (Figure 3) the following T-dependencies are obtained:

$$k(T, AH_2) = (7.3 \pm 0.2) \cdot 10^{10} \cdot \exp\left(\frac{-1040 \pm 140\ K}{T}\right) /L\ mol^{-1}\ s^{-1} \quad (9)$$

$$k(T, AH^{-}) = (1.8 \pm 0.1) \cdot 10^{11} \cdot \exp\left(\frac{-1200 \pm 240\,K}{T}\right) /L\,mol^{-1}\,s^{-1} \quad (10)$$

$$k(T, A^{2-}) = (1.4 \pm 0.1) \cdot 10^{12} \cdot \exp\left(\frac{-1830 \pm 110\,K}{T}\right) /L\,mol^{-1}\,s^{-1} \quad (11)$$

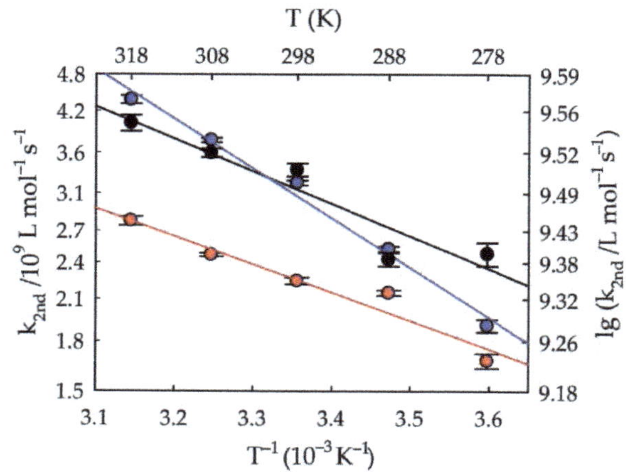

Figure 3. Arrhenius plot of pH dependent OH radical reactions with pimelic acid in their different protonation forms AH_2 (red), AH^- (black), and A^{2-} (blue).

Similar to the above-mentioned succinic acid the T dependency of the OH radical reaction of pimelic acid at pH = 8 is strongly dependent on T. The monoanionic form of both acids exhibits a similar T-dependency like the fully protonated acid with a faster reaction rate.

3.1. Oxidation Mechanism

The general reaction mechanism of OH radicals with organic compounds (dicarboxylic acids) in the aqueous media can be described by an H atom abstraction mechanism. The OH radical attack occurs predominantly at the weakest C-H bond of the molecule [29,36–38]. In general hydrogen bond dissociation energies (BDEs) of a dicarboxylic acid show the following trend: $CH < CH_2 < CH_3 < OH$ [39]. The BDE of the C-H bond was estimated to be 395 kJ mol^{-1} [37] and 397.8 ± 5.1 kJ mol^{-1} [40] for succinic acid. An alkyl radical is formed after the H atom abstraction which reacts in the presence of O_2 to a peroxyl radical [41]. Depending on the molecule structure, the derived peroxyl radical can recombine to either form a tetroxide or react further in a unimolecular decay process [42]. In the case of succinic acid (C_4 dicarboxylic acid), the oxidation leads to the formation of malic acid and oxaloacetic acid via the Russell mechanism [16,43], oxaloacetic acid and hydrogen peroxide via Bennett and Summers mechanism [44], or by the fragmentation of the tetroxide which results in oxygen and an alkoxy radical [42]. These alkoxy radicals can react in a subsequent step to a CO_2^- radical and 3-oxopropanoic acid [15]. A second possibility is the intramolecular H-shift of the alkoxy radical, which leads in a subsequent reaction to oxaloacetic acid and a hydroperoxyl radical (HO_2) [42]. The alkoxy radical might also react with another organic molecule and undergo an H atom abstraction reaction to form an alkyl radical and malic acid, while the contribution for this process is small. Hence, this reaction plays a role only in highly concentrated solutions, as they can occur in aerosols. In addition, the tetroxide decays by disproportionation in the solvent cage to yield 3-hydroxypropionic acid proposed by Alterie et al. [11]. In a study using aerosol flow tube experiments from Chan et al. [17] it was reported that the oxidation products of succinic acid are mainly C_4-compounds, but fragmentation species such as oxomalonic acid (C_3) and glyoxylic acid (C_2) were also found.

Similarly, pimelic acid oxidation by OH radicals takes place at the α-position and leads to the main reaction products 2-hydroxyheptanedioic acid and 2-oxoheptanedioic acid via the Russell mechanism and chain shortening products such as 6-oxohexanoic acid.

3.2. Activation Parameters

The Gibbs free energy of activation, enthalpy of activation, and entropy of activation were derived as follows: the Gibbs free energy of activation, $\Delta G^{\ddagger}$, was calculated using the Eyring equation [35] (Equation (12)). The parameter k_B and h depict the Boltzmann's constant and Planck's constant

$$k(T) = \frac{k_B T}{h} \cdot e^{(-\Delta G^{\ddagger}/RT)} \ /\mathrm{L\ mol^{-1}\ s^{-1}} \quad (12)$$

The enthalpy of activation, $\Delta H^{\ddagger}$, is related to the activation energy, E_A, as follows in Equation (13)

$$\Delta H^{\ddagger} = E_A - RT \ /\mathrm{kJ\ mol^{-1}} \quad (13)$$

The entropy of activation, $\Delta S^{\ddagger}$, was derived from the pre-exponential factor, A, based on Equation (14)

$$\Delta S^{\ddagger} = R\left(\ln A - \ln\left(\frac{k_B T}{h}\right) - 1\right) /\mathrm{J\ K^{-1}\ mol^{-1}} \quad (14)$$

The activation parameters derived for the investigated OH radical oxidation reactions are summarized in Table 3.

Table 3. Activation parameters of the studied OH reactions in aqueous solution. Arrhenius pre-exponential factor, A, and activation energy, E_A, Gibbs free energy, $\Delta G^{\ddagger}$, enthalpy, $\Delta H^{\ddagger}$, and entropy, $\Delta S^{\ddagger}$, of activation.

Compounds	A/L mol^{-1} s^{-1}	E_A/kJ mol^{-1}	$\Delta H^{\ddagger}$/kJ mol^{-1}	$\Delta S^{\ddagger}$/J K^{-1} mol^{-1}	$\Delta G^{\ddagger}$/kJ mol^{-1}
			Succinic acid		
Diacid	$(2.1 \pm 0.1) \cdot 10^{10}$	13 ± 2	10 ± 2	$-(56 \pm 3)$	27 ± 7
Monoanion	$(1.5 \pm 0.1) \cdot 10^{10}$	9 ± 3	7 ± 3	$-(57 \pm 4)$	24 ± 11
Dianion	$(2.9 \pm 0.2) \cdot 10^{11}$	15 ± 3	13 ± 3	$-(34 \pm 2)$	23 ± 7
			Pimelic acid		
Diacid	$(7.3 \pm 0.2) \cdot 10^{10}$	9 ± 1	6 ± 1	$-(45 \pm 1)$	20 ± 4
Monoanion	$(1.8 \pm 0.1) \cdot 10^{11}$	10 ± 2	8 ± 2	$-(38 \pm 2)$	19 ± 6
Dianion	$(1.4 \pm 0.1) \cdot 10^{12}$	15 ± 1	13 ± 1	$-(21 \pm 1)$	19 ± 2

The value of the activation energy, E_A, slightly increase by changing the pH value as well as the derived pre-exponential factors, A, increases. Positive values for enthalpy, $\Delta H^{\ddagger}$, in the range from 6 to 10 kJ mol^{-1} for acidic conditions and 13 kJ mol^{-1} for alkaline conditions were obtained. For both reactants, the negative values of the entropy, $\Delta S^{\ddagger}$, increase to less negative values by increasing the pH, while the pH-dependence of pimelic acid is stronger than that of succinic acid. In contrast to $\Delta H^{\ddagger}$, the calculated values for the free Gibbs enthalpy, $\Delta G^{\ddagger}$, do not show a strong pH-dependency. The monoanionic form of succinic acid has a smaller pre-exponential factor and a lower activation energy than the fully protonated or deprotonated form. In contrast, the monoanionic speciation of pimelic acid lies exactly between these two forms, which might be explained by the larger contribution of the H atom abstraction. The obtained values of the activation parameters are comparable with the literature values for the H-atom abstraction reactions of short-chain carbon compounds such as acids, dicarboxylic acids, diols, and ketones [14,28,29,45].

The obtained room temperature (T = 298 K) rate constants from the present study are compared with the available literature data in Table 4.

Table 4. Second-order kinetic rate constants for the reactions of the studied dicarboxylic acids with OH in aqueous solution. Bold printing corresponds to data from the present study.

Compounds	k_{OH}/L mol^{-1} s^{-1}	Remarks	Reference
		Succinic acid	
AH_2	$1.0 \cdot 10^{8}$ [a]	pH = 2, T = not reported/room temperature, PR-comp. kinetics, ref. thymine (slope 0.023)	[46]
	$1.33 \cdot 10^{8}$ [a]	pH = 2.0–2.2, T = 303.15 K, Fenton type-comp. kinetics, ref. methanol (slope 0.097)	[24]
	$(3.1 \pm 0.3) \cdot 10^{8}$ [a]	pH = 1.5–10, T = 298.25 K, PR-build-up kinetics at λ = 250–350 nm, (pK_a 4.16, 5.61)	[47]
	$(1.1 \pm 0.1) \cdot 10^{8}$ [a, b] $(1.06 \pm 0.1) \cdot 10^{8}$ [c]	pH = 1.5, T = 298.15 K, PR-comp. kinetics, ref. SCN^- (slope 0.0089), T-dependent measurement, (pK_a 4.16, 5.61)	[19][d]
	$1.1 \cdot 10^{8}$	Estimation method	[48]
	$7.7 \cdot 10^{7}$	Estimation method	[49]
	$\mathbf{(1.3 \pm 0.2) \cdot 10^{8}}$ [b]	**pH = 2, T = 298.15 K**	**This work**
AH^-	$2.0 \cdot 10^{8}$	Estimation method	[49]
	$\mathbf{(5.1 \pm 0.1) \cdot 10^{8}}$ [b]	**pH = 4.9, T = 298.15 K**	**This work**
A^{2-}	$(5.3 \pm 0.5) \cdot 10^{8}$ [a]	pH = 9, borax buffer, T = 293.15 K, FP-comp. kinetics, ref. 4-ferrocenyl-butanoate (slope 0.049)	[21]
	$(5.0 \pm 0.5) \cdot 10^{8}$ [a, b] $(4.8 \pm 0.48) \cdot 10^{8}$ [c]	pH = 7.8, T = 298.15 K, LFP-comp. kinetics, ref. SCN^- (slope 0.0403), T-dependent measurement, (pK_a 4.16, 5.61)	[19]
	$4.6 \cdot 10^{8}$	Estimation method	[49]
	$\mathbf{(5.3 \pm 0.2) \cdot 10^{8}}$ [b]	**pH = 8, T = 298.15 K**	**This work**
		Pimelic acid	
AH_2	$2.4 \cdot 10^{9}$ [a] $3.5 \cdot 10^{9}$ [e]	pH = 2, T = not reported/room temperature, PR-comp. kinetics, ref. thymine (slope 0.558)	[46]
	$(2.4 \pm 0.7) \cdot 10^{9}$	pH = 2, T = 298.15 K, LFP-build-up kinetics at λ = 244 nm	[29]
	$4.3 \cdot 10^{9}$	Estimation method	[48]
	$1.6 \cdot 10^{9}$	Estimation method	[49]
	$\mathbf{(2.2 \pm 0.1) \cdot 10^{9}}$ [b]	**pH = 2, T = 298.15 K**	**This work**
AH^-	$2.3 \cdot 10^{9}$	Estimation method	[49]
	$\mathbf{(3.4 \pm 0.1) \cdot 10^{9}}$ [b]	**pH = 4.6, T = 298.15 K**	**This work**
A^{2-}	$(2.9 \pm 0.8) \cdot 10^{9}$	pH = 9, T = 298.15 K, LFP-build-up kinetics at λ = 244 nm	[29]
	$3.0 \cdot 10^{9}$	Estimation method	[49]
	$\mathbf{(3.2 \pm 0.1) \cdot 10^{9}}$ [b]	**pH = 8, T = 298.15 K**	**This work**

LFP—laser flash photolysis, FP—Flash photolysis, PR—pulse radiolysis, [a] Reported in the publication, [b] reference constant k(OH + SCN^-) = $1.24 \cdot 10^{10}$ L mol^{-1} s^{-1} [50], [c] reference constant k(OH + SCN^-) = $1.19 \cdot 10^{10}$ L mol^{-1} s^{-1} [27], [d] in the original publication by Ervens et al. [19] the values of the rate constant of the fully protonated and fully deprotonated form of succinic in Data Table 11 of Ervens et al., [19] were confused with each other, while the expressions in the abstract are correct. [e] NIST Solution Kinetics Database [51].

Most of the rate constants for succinic acid were measured by relative-rate techniques, with the exception of the study by Cabelli and Bielski [47]. In this study, the authors used the product build-up kinetic method at λ = 250–330 nm and reported a pH-independent rate constant k(298 K) = $(3.1 \pm 0.3) \cdot 10^{8}$ L mol^{-1} s^{-1}. These findings are in contradiction to the observed rate constants for succinic acid and other carboxylic acids [9,14,20–22,29,46,52]. Ervens et al. also described the oxidation of succinic acid by OH radicals, but in this study the monoanionic form was not investigated [19]. It should be mentioned that a source of uncertainty in the rate constants could be the use of different reference constants (cf. Table 4). For example, in the study by Ervens et al. [19] a slightly larger reference constant was used. Correcting the given rate constants for succinic acid with the reference constant used in this study by Zhu et al. [27] results in smaller rate constants $k(AH_2) = (1.06 \pm 0.1) \cdot 10^{8}$ L mol^{-1} s^{-1} and $k(A^{2-}) = (4.8 \pm 0.48) \cdot 10^{8}$ L mol^{-1} s^{-1}. Chan et al. [17] reported an OH radical rate constant k = (7.4 ± 2.4) · 10^7 L mol^{-1} s^{-1} by analyzing succinic acid decomposition with DART mass spectrometry in their aerosol flow tube experiments. Under the conditions used, most succinic acid should be present in the fully protonated form with small contributions of the monoanionic form and fits quiet well to the received rate constants from this study and reported rate constants in Table 4 [17]. However, the OH radical rate constants of succinic acid obtained from this study agree well with the reported values for the low pH range (cf. Table 5). Minakata et al. [48] or Doussin and Monod [49] developed two prediction methods for OH radical rate constants in aqueous solution based on structure activity relationships (SAR) (see more details in the Supplementary Materials). The obtained results for pH = 2 are consistent with the rate constant measured in this study. The method by Doussin and Monod [49] also reflects the pH effect on the rate constant for both acids, since it includes different group increments for carboxyl-/carboxylate group. The results of the OH radical reaction with pimelic acid are comparable with measurements using a build-up kinetic method [29] as well as relative-rate technique [46].

In the following Figure 4, the second-order OH radical rate constants are plotted against the carbon chain length for the different speciation of various mono- and dicarboxylic acids (cf. Tables S1 and S2). As expected, a significant increase in the reaction rate is shown in Figure 4 by increasing chain length of the dicarboxylic acids. This is because the length of the alkene backbone influences the number of abstractable hydrogen atoms and so the probability of an effective collision increases. Furthermore, the steric hindrance by the terminal carboxyl groups also contributes to these results, since the reaction rate of the OH radical with the carboxylic acid group is lower compared to the CH_2 group. This behavior is expressed in the estimation methods by the different fragments of the respective functional group [48,49].

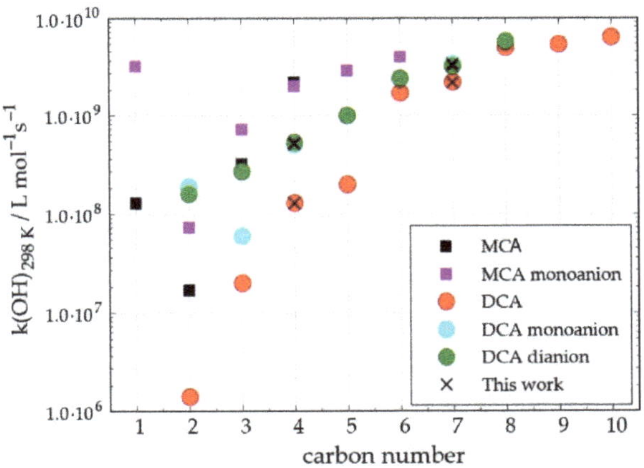

Figure 4. OH radical second order rate constants k_{2nd} as a function of the carbon chain length of the carboxylic acids (see more details in the Supplementary Materials).

The pH dependence of the OH radical reaction of dicarboxylic acids could be explained by the following effects: (i) the increasing contribution of an electron transfer reaction between the carboxylate group and the OH radical, (ii) an increased attraction between the carboxylate group and the electrophilic OH radical by electrostatic interactions, (iii) the reduction of the bond dissociation energy (BDE) of the C-H bond by deprotonation of the carboxyl function, and (iv) the formation of a ring-shaped molecular structure by intramolecular interactions of the carboxylate group with the CH_2 group. The formation of a ring-shaped molecular structure can be neglected, since significantly greater effects can be seen with the shorter dicarboxylic acid. Even the influence of the deprotonation of the carboxyl function on the BDE of the adjacent C-H bond cannot explain the change by one order of magnitude, since the change of the BDE is too small. According to the data summarized in

Hoffmann et al. [37], the change in BDE must be more than 10 kJ mol^{-1} in order for the OH radical rate constant to increase by more than one order of magnitude. Hence, the increase of the OH radical rate constant could be explained by a greater attraction leading to an increase in the number of molecular collisions or, more likely, by the contribution of an electron transfer reaction. If the carboxyl group is deprotonated, the electron transfer reaction can take place, this reaction pathway is inhibited by the protonation of the carboxyl group and the H atom abstraction from the OH group is a very slow process due to the higher BDE, which leads to a reduction of the rate constant. This should be clarified by further studies using ab initio theoretical methods.

4. Atmospheric Implications

It was the main objective of the present study to clarify the influence of the pH and temperature on the rate constant for the reactions of OH with succinic and pimelic acid in their respective three forms.

The performed kinetic measurements allow us to estimate the lifetime of the studied dicarboxylic acids in the atmospheric aqueous phase. Atmospheric aqueous phase lifetimes (τ) were calculated in this context by the following simple equation (Equation (15)).

$$\tau = (k \cdot [OH]_0)^{-1} / s \tag{15}$$

The following OH model concentrations were used for different scenarios: an urban cloud case of $3.5 \cdot 10^{-15}$ mol L^{-1}, a remote cloud case of $2.2 \cdot 10^{-14}$ mol L^{-1}, an urban aerosol case of $4.4 \cdot 10^{-13}$ mol L^{-1}, and a remote aerosol case of $3.0 \cdot 10^{-12}$ mol L^{-1} [14]. Based on the assumption that the OH radicals present reacts exclusively with the dicarboxylic acid, the following lifetimes in the aqueous phase were obtained, using the rate constants at T = 278 K and T = 298 K to illustrate the atmospheric conditions (cf. Table 5).

Table 5. Atmospheric aqueous-phase lifetimes calculated for urban and remote areas.

Compounds	Cloud Case Urban $[OH] = 3.5 \cdot 10^{-15}$ /mol L^{-1}	Cloud Case Remote $[OH] = 2.2 \cdot 10^{-14}$ /mol L^{-1}	Aerosol Case Urban $[OH] = 4.4 \cdot 10^{-13}$ /mol L^{-1}	Aerosol Case Remote $[OH] = 3.0 \cdot 10^{-12}$ /mol L^{-1}
		Succinic acid		
AH_2	25.4–41.9 d	4.0–6.7 d	4.9–8.0 h	43–70 min
AH^-	6.2–7.7 d	23.8–29.4 h	72–88 min	11–13 min
A^{2-}	6.5–11 d	24.8–42.1 h	74–126 min	11–19 min
		Pimelic acid		
AH_2	36.1–46.7 h	5.7–7.4 h	17–22 min	2–3 min
AH^-	24.8–41.8 h	3.9–6.6 h	12–20 min	2–3 min
A^{2-}	24.8–41.8 h	3.9–6.6 h	12–20 min	2–3 min

Depending on the scenario and the used temperature, the calculated atmospheric lifetimes can range from a few minutes to a few days. The lifetime in aerosols is, as expected, considerably shorter than in clouds (Table 5). The tropospheric water phase can occur at both low and very high pH values. Both concentrated deliquescent particles and diluted cloud droplets can have a wide pH range from strongly acidic to slightly alkaline [9]. Therefore, all three species of the dicarboxylic acids AH_2, AH^-, and A^{2-} can play a role in the tropospheric oxidation process of the investigated DCAs (cf. Figure 1). In urban and continental aerosol particles with a lower pH from zero to five mainly the fully protonated form AH_2 is present. The monoanionic form of both dicarboxylic acid occurs mainly in moderate tropospheric pH regimes from 3.0 to 6.5 as found in cloud droplets or rain. Finally, the fully deprotonated form could be present in marine aerosols, haze, and fog with elevated pH values up to nine.

Figure 5 shows the overall rate constant for T = 278 K and 298 K as function of the pH. The overall rate constant was calculated by using the Equation (2), the acid–base equilibrium contributions, and the individual rate constants of the respective speciation of the dicarboxylic acids. The OH radical oxidation generally results in further oxygenated compounds, which then can undergo further subsequent accretion reactions such as aldol, acetal, and ester formation. The before-mentioned accretion processes can become more important when droplet evaporation occurs [9,53], due to the presence of more oxygenated dicarboxylic acid or smaller dicarboxylic acids such as malic acid and oxaloacetic acid from the OH oxidation process and the water loss in the droplet. In addition, the lifetime of an aerosol particle can be estimated at about 5 days, whereas the lifetime of a cloud droplet is shorter than 18 h. This means that the tropospheric lifetime of dicarboxylic acids corresponds more to the lifetime of the aerosol phase. In addition, because of the different reactivity of dicarboxylic acids, succinic acid has a longer lifetime than pimelic acid, which explains the findings of Teich et al. [7]. Since pH dependence occurs during the oxidation of dicarboxylic acids, pH under cloud droplet conditions plays an important role, whereas due to the high acidity in aerosol particles the pH influence plays a minor role.

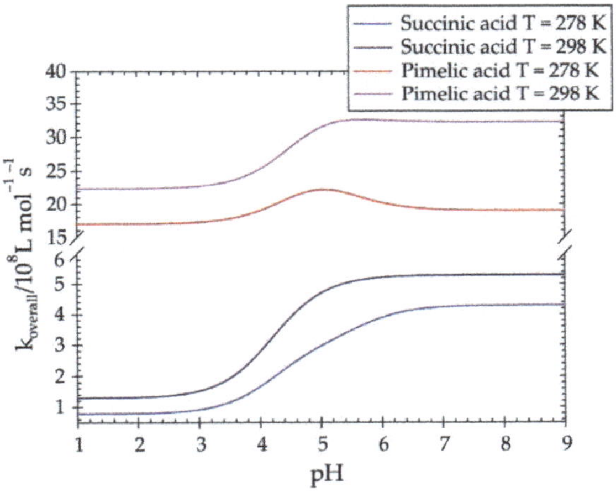

Figure 5. Overall OH radical rate constant ($k_{overall}$) as function of the pH at T = 278 K and T = 298 K.

5. Conclusions

Within the present study, the pH- and temperature-dependent oxidation reactions of succinic acid and pimelic acid were investigated to determine the OH radical rate constants. The temperature-dependency of the acid-base equilibria of the dicarboxylic acids shows no significant influence of the ratio of the different species AH_2, AH^-, or A^{2-}. Moreover, the internal absorption effect of the used compounds can also be neglected, since the molar attenuation coefficient is too small for a significant contribution under the used conditions. The internal absorption effect can occur when using the competition kinetics, leading to a virtual increase of the measured rate constant, but this effect can be neglected for the two dicarboxylic acids under the conditions used. The following behavior of the OH radical rate constants was observed: the rate constant increases with increasing pH, due to the deprotonation of the carboxylic acid group of the molecule and thus the possible contribution of an electron transfer reaction increases. This effect is much stronger with small dicarboxylic acids than with long-chain dicarboxylic acids. The results obtained were compared with SAR prediction tools and revealed a very good agreement between the measured and the estimated values. Furthermore, the tropospheric lifetime of dicarboxylic acids was estimated to range from a few minutes under aerosol conditions to several hours under cloud conditions. This oxidation reaction by OH radicals can act as a sink for dicarboxylic acids, as well as a source of smaller dicarboxylic acids or multifunctional compounds.

Supplementary Materials: The following are available online at http://www.mdpi.com/2073-4433/11/4/320/s1, Figure S1. Measured absorbance-time profiles of the radical species $(SCN)_2^-$, Figure S2. Ratios of the absorbance of $(SCN)_2^-$, calculation of the estimated rate constants, Table S1. OH radical reaction rate constants (k_{OH}(298 K) in L mol^{-1} s^{-1})) of monocarboxylic acids., Table S2. OH radical reaction rate constants (k_{OH}(298 K) in L mol^{-1} s^{-1})) of dicarboxylic acids.

Author Contributions: Conceptualization, T.S.; investigation, L.W., A.E. and J.M.; writing—original draft preparation, T.S.; writing—review and editing, T.S. and H.H.; supervision, T.S. All authors have read and agreed to the published version of the manuscript.

Funding: This research received no external funding.

Acknowledgments: The authors thank Kornelia Pielok and Axel Rohmer for technical assistance.

Conflicts of Interest: The authors declare no conflict of interest.

References

1. Compernolle, S.; Müller, J.F. Henry's law constants of diacids and hydroxy polyacids: Recommended values. *Atmos. Chem. Phys.* **2014**, *14*, 2699–2712. [CrossRef]
2. Sander, R. Compilation of Henry's law constants (version 4.0) for water as solvent. *Atmos. Chem. Phys.* **2015**, *15*, 4399–4981. [CrossRef]
3. Chebbi, A.; Carlier, P. Carboxylic acids in the troposphere, occurrence, sources, and sinks: A review. *Atmos. Environ.* **1996**, *30*, 4233–4249. [CrossRef]
4. Kawamura, K.; Bikkina, S. A review of dicarboxylic acids and related compounds in atmospheric aerosols: Molecular distributions, sources and transformation. *Atmos. Res.* **2016**, *170*, 140–160. [CrossRef]
5. Kawamura, K.; Kasukabe, H.; Barrie, L.A. Secondary formation of water-soluble organic acids and α-dicarbonyls and their contributions to total carbon and water-soluble organic carbon: Photochemical aging of organic aerosols in the Arctic spring. *J. Geophys. Res. Atmos.* **2010**, *115*. [CrossRef]
6. Kawamura, K.; Seméré, R.; Imai, Y.; Fujii, Y.; Hayashi, M. Water soluble dicarboxylic acids and related compounds in Antarctic aerosols. *J. Geophys. Res. Atmos.* **1996**, *101*, 18721–18728. [CrossRef]
7. Teich, M.; van Pinxteren, D.; Herrmann, H. A one year study of functionalised medium-chain carboxylic acids in atmospheric particles at a rural site in Germany revealing seasonal trends and possible sources. *J. Atmos. Chem.* **2019**, *76*, 115–132. [CrossRef]
8. Van Pinxteren, D.; Neusüß, C.; Herrmann, H. On the abundance and source contributions of dicarboxylic acids in size-resolved aerosol particles at continental sites in central Europe. *Atmos. Chem. Phys.* **2014**, *14*, 3913–3928. [CrossRef]
9. Herrmann, H.; Schaefer, T.; Tilgner, A.; Styler, S.A.; Weller, C.; Teich, M.; Otto, T. Tropospheric aqueous-phase chemistry: Kinetics, mechanisms, and its coupling to a changing gas phase. *Chem. Rev.* **2015**, *115*, 4259–4334. [CrossRef]
10. Seinfeld, J.H.; Pandis, S.N. *Atmospheric Chemistry and Physics: From Air Pollution to Climate Change*; Wiley: Hoboken, NJ, USA, 2006.
11. Altieri, K.E.; Seitzinger, S.P.; Carlton, A.G.; Turpin, B.J.; Klein, G.C.; Marshall, A.G. Oligomers formed through in-cloud methylglyoxal reactions: Chemical composition, properties, and mechanisms investigated by ultra-high resolution FT-ICR mass spectrometry. *Atmos. Environ.* **2008**, *42*, 1476–1490. [CrossRef]
12. Liu, T.; Clegg, S.L.; Abbatt, J.P.D. Fast oxidation of sulfur dioxide by hydrogen peroxide in deliquesced aerosol particles. *Proc. Natl. Acad. Sci. USA* **2020**, *117*, 1354. [CrossRef] [PubMed]
13. Hazra, M.K.; Sinha, A. Formic acid catalyzed hydrolysis of SO_3 in the gas phase: A barrierless mechanism for sulfuric acid production of potential atmospheric importance. *J. Am. Chem. Soc.* **2011**, *133*, 17444–17453. [CrossRef] [PubMed]
14. Herrmann, H.; Hoffmann, D.; Schaefer, T.; Bräuer, P.; Tilgner, A. Tropospheric aqueous-phase free-radical chemistry: Radical sources, spectra, reaction kinetics and prediction tools. *ChemPhysChem* **2010**, *11*, 3796–3822. [CrossRef] [PubMed]

15. Vel Leitner, N.K.; Dore, M. Hydroxyl radical induced decomposition of aliphatic acids in oxygenated and deoxygenated aqueous solutions. *J. Photochem. Photobiol. A* **1996**, *99*, 137–143. [CrossRef]
16. Gao, S.S.; Abbatt, J.P.D. Kinetics and mechanism of OH oxidation of small organic dicarboxylic acids in ice: Comparison to behavior in aqueous solution. *J. Phys. Chem. A* **2011**, *115*, 9977–9986. [CrossRef]
17. Chan, M.N.; Zhang, H.; Goldstein, A.H.; Wilson, K.R. Role of water and phase in the heterogeneous oxidation of solid and aqueous succinic acid aerosol by hydroxyl radicals. *J. Phys. Chem. C* **2014**, *118*, 28978–28992. [CrossRef]
18. Chin, M.; Wine, P.H. *A Temperature-Dependent Competitive Kinetics Study of the Aqueous-Phase Reactions of OH Radicals with Formate, Formic Acid, Acetate, Acetic Acid, and Hydrated Formaldehyde*; Lewis: Hangzhou, China, 1994; pp. 85–96. [CrossRef]
19. Ervens, B.; Gligorovski, S.; Herrmann, H. Temperature-dependent rate constants for hydroxyl radical reactions with organic compounds in aqueous solutions. *Phys. Chem. Chem. Phys.* **2003**, *5*, 1811–1824. [CrossRef]
20. Gligorovski, S.; Rousse, D.; George, C.H.; Herrmann, H. Rate constants for the OH reactions with Oxygenated organic compounds in aqueous solution. *Int. J. Chem. Kinet.* **2009**, *41*, 309–326. [CrossRef]
21. Logan, S.R. Redox reactions of organic radicals with ferrocene/ferricenium species in aqueous solution. Part 1. Radicals derived from carboxylic acids. *J. Chem. Soc. Perkin Trans. 2* **1989**, 751–754. [CrossRef]
22. Martin, L.R.; Mezyk, S.P.; Mincher, B.J. Determination of arrhenius and thermodynamic parameters for the aqueous reaction of the hydroxyl radical with lactic acid. *J. Phys. Chem. A* **2009**, *113*, 141–145. [CrossRef]
23. Otto, T.; Schaefer, T.; Herrmann, H. Aqueous-phase oxidation of terpene-derived acids by atmospherically relevant radicals. *J. Phys. Chem. A* **2018**, *122*, 9233–9241. [CrossRef] [PubMed]
24. Walling, C.; El-Taliawi, G.M. Fenton's reagent. II. Reactions of carbonyl compounds and alpha, beta-unsaturated acids. *J. Am. Chem. Soc.* **1973**, *95*, 844–847. [CrossRef]
25. Adler, H.; Sirén, H. Study on dicarboxylic acids in aerosol samples with capillary electrophoresis. *J. Anal. Methods Chem.* **2014**, *2014*, 498168. [CrossRef] [PubMed]
26. Martinsson, J.; Monteil, G.; Sporre, M.K.; Kaldal Hansen, A.M.; Kristensson, A.; Eriksson Stenström, K.; Swietlicki, E.; Glasius, M. Exploring sources of biogenic secondary organic aerosol compounds using chemical analysis and the FLEXPART model. *Atmos. Chem. Phys.* **2017**, *17*, 11025–11040. [CrossRef]
27. Zhu, L.; Nicovich, J.M.; Wine, P.H. Temperature-dependent kinetics studies of aqueous phase reactions of hydroxyl radicals with dimethylsulfoxide, dimethylsulfone, and methanesulfonate. *Aquat. Sci.* **2003**, *65*, 425–435. [CrossRef]
28. Schaefer, T.; Herrmann, H. Competition kinetics of OH radical reactions with oxygenated organic compounds in aqueous solution: Rate constants and internal optical absorption effects. *Phys. Chem. Chem. Phys.* **2018**, *20*, 10939–10948. [CrossRef]
29. Herrmann, H. Kinetics of aqueous phase reactions relevant for atmospheric chemistry. *Chem. Rev.* **2003**, *103*, 4691–4716. [CrossRef]
30. Martell, A.E.; Smith, R.M. *Critical Stability Constants*; Plenum Press: New York, NY, USA, 1976; pp. 1–4. [CrossRef]
31. Von Smoluchowski, M. Versuch einer mathematischen theorie der koagulations kinetic kolloider lösungen. *Z. Phys. Chem.* **1917**, *92*, 129. [CrossRef]
32. He, L.; Schaefer, T.; Otto, T.; Kroflič, A.; Herrmann, H. Kinetic and theoretical study of the atmospheric aqueous-phase reactions of OH radicals with methoxyphenolic compounds. *J. Phys. Chem. A* **2019**, *123*, 7828–7838. [CrossRef]
33. Kroflič, A.; Schaefer, T.; Huš, M.; Phuoc Le, H.; Otto, T.; Herrmann, H. OH radicals reactivity towards phenol-related pollutants in water: Temperature dependence of the rate constants and novel insights into the [OH–phenol] adduct formation. *Phys. Chem. Chem. Phys.* **2020**, *22*, 1324–1332. [CrossRef]
34. Otto, T.; Schaefer, T.; Herrmann, H. Aqueous-phase oxidation of cis-β-Isoprene epoxydiol by hydroxyl radicals and its impact on atmospheric isoprene processing. *J. Phys. Chem. A* **2019**, *123*, 10599–10608. [CrossRef] [PubMed]
35. Atkins, P.; De Paula, J. *Physical Chemistry*, 8th ed.; W. H. Freeman and Company: New York, NY, USA, 2006.
36. Asmus, K.D.; Moeckel, H.; Henglein, A. Pulse radiolytic study of the site of hydroxyl radical attack on aliphatic alcohols in aqueous solution. *J. Phys. Chem.* **1973**, *77*, 1218–1221. [CrossRef]

37. Hoffmann, D.; Weigert, B.; Barzaghi, P.; Herrmann, H. Reactivity of poly-alcohols towards OH, NO_3 and SO_4^- in aqueous solution. *Phys. Chem. Chem. Phys.* **2009**, *11*, 9351–9363. [CrossRef] [PubMed]
38. Buxton, G.V.; Greenstock, C.L.; Helman, W.P.; Ross, A.B. Critical review of rate constants for reactions of hydrated electrons, hydrogen atoms and hydroxyl radicals (OH/O^-) in aqueous solution. *J. Phys. Chem. Ref. Data* **1988**, *17*, 513–886. [CrossRef]
39. Luo, Y.R. *Handbook of Bond Dissociation Energies in Organic Compounds*; CRC Press: Boca Raton, FL, USA, 2002. [CrossRef]
40. Denisov, E.T.; Tumanov, V.E. Estimation of the bond dissociation energies from the kinetic characteristics of liquid-phase radical reactions. *Russ. Chem. Rev.* **2005**, *74*, 825–858. [CrossRef]
41. Neta, P.; Huie, R.E.; Ross, A.B. Rate constants for reactions of peroxyl radicals in fluid solutions. *J. Phys. Chem. Ref. Data* **1990**, *19*, 413–513. [CrossRef]
42. Von Sonntag, C.; Schuchmann, H.-P. The Elucidation of peroxyl radical reactions in aqueous solution with the help of radiation-chemical methods. *Angew. Chem. Int. Ed. Engl.* **1991**, *30*, 1229–1253. [CrossRef]
43. Russell, G.A. Deuterium-isotope effects in the autoxidation of Aralkyl hydrocarbons. Mechanism of the interaction of Peroxy Radicals1. *J. Am. Chem. Soc.* **1957**, *79*, 3871–3877. [CrossRef]
44. Bennett, J.E.; Summers, R. Product studies of the mutual termination reactions of sec-alkylperoxy radicals: Evidence for non-cyclic termination. *Can. J. Chem.* **1974**, *52*, 1377–1379. [CrossRef]
45. Schaefer, T.; van Pinxteren, D.; Herrmann, H. Multiphase chemistry of glyoxal: Revised kinetics of the alkyl radical reaction with molecular oxygen and the reaction of glyoxal with OH, NO_3, and SO_4^- in aqueous solution. *Environ. Sci. Technol.* **2015**, *49*, 343–350. [CrossRef]
46. Scholes, G.; Willson, R.L. γ-Radiolysis of aqueous thymine solutions. Determination of relative reaction rates of OH radicals. *Trans. Faraday Soc.* **1967**, *63*, 2983–2993. [CrossRef]
47. Cabelli Diane, E.; Bielski Benon, H.J. A pulse radiolysis study of some dicarboxylic acids of the citric acid cycle. The kinetics and spectral properties of the free radicals formed by reaction with the OH radical. *Z. Naturforsch. B Chem. Sci.* **1985**, *40*, 1731. [CrossRef]
48. Minakata, D.; Li, K.; Westerhoff, P.; Crittenden, J. Development of a group contribution method to predict aqueous phase hydroxyl radical (HO•) reaction rate constants. *Environ. Sci. Technol.* **2009**, *43*, 6220–6227. [CrossRef] [PubMed]
49. Doussin, J.F.; Monod, A. Structure-activity relationship for the estimation of OH-oxidation rate constants of carbonyl compounds in the aqueous phase. *Atmos. Chem. Phys.* **2013**, *13*, 11625–11641. [CrossRef]
50. Chin, M.; Wine, P.H. A temperature-dependent kinetics study of the aqueous phase reactions $OH + SCN^- \rightarrow SCNOH^-$ and $SCN + SCN^- \rightleftarrows (SCN)_2^-$. *J. Photochem. Photobiol. A* **1992**, *69*, 17–25. [CrossRef]
51. NIST. *Solution Kinetics Database Version 3.0*; NIST: Gaithersburg, MD, USA, 1998.
52. Vel Leitner, N.K.; Doré, M. Mecanisme d'action des radicaux OH sur les acides glycolique, glyoxylique, acetique et oxalique en solution aqueuse: Incidence sur la consammation de peroxyde d'hydrogene dans les systemes H_2O_2UV et $O_3H_2O_2$. *Water Res.* **1997**, *31*, 1383–1397. [CrossRef]
53. Barker, J.R.; Steiner, A.L.; Wallington, T.J. *Advances in Atmospheric Chemistry*; World Scientific: Singapore, 2017; Volume 1, p. 608. [CrossRef]

atmosphere

MDPI

Article

Modeling of Carbonyl/Ammonium Sulfate Aqueous Brown Carbon Chemistry via UV/Vis Spectral Decomposition

Mengjie Fan †, Shiqing Ma, Nahin Ferdousi, Ziwei Dai ‡ and Joseph L. Woo *

Department of Chemical and Biomolecular Engineering, Lafayette College, Easton, PA 18042, USA; fanmj@seas.upenn.edu (M.F.); mas@lafayette.edu (S.M.); ferdousn19@gmail.com (N.F.); zdai2@nd.edu (Z.D.)

* Correspondence: wooj@lafayette.edu; Tel.: +1-610-330-3291

† Current address: Department of Chemical and Biomolecular Engineering, University of Pennsylvania, Philadelphia, PA 19104, USA.

‡ Current address: Department of Chemical and Biomolecular Engineering, University of Notre Dame, Notre Dame, IN 46556, USA.

Received: 30 January 2020; Accepted: 3 April 2020; Published: 7 April 2020

Abstract: The proper characterization of aqueous brown carbon (BrC) species, their formation, and their light absorbance properties is critical to understanding the aggregate effect that they have on overall atmospheric aerosol climate forcing. The contribution of dark chemistry secondary organic aerosol (SOA) products from carbonyl-containing organic compounds (CVOCs) to overall aqueous aerosol optical properties is expected to be significant. However, the multiple, parallel pathways that take place within CVOC reaction systems and the differing chromophoricity of individual products complicates the ability to reliably model the chemical kinetics taking place. Here, we proposed an alternative method of representing UV-visible absorbance spectra as a composite of Gaussian lineshape functions to infer kinetic information. Multiple numbers of curves and different CVOC/ammonium reaction systems were compared. A model using three fitted Gaussian curves with magnitudes following first-order kinetics achieved an accuracy within 65.5% in the 205–300-nm range across multiple organic types and solution aging times. Asymmetrical peaks that occurred in low-200-nm wavelengths were decomposed into two overlapping Gaussian curves, which may have been attributable to different functional groups or families of reaction products. Component curves within overall spectra exhibited different dynamics, implying that the utilization of absorbance at a single reference wavelength to infer reaction rate constants may result in misrepresentative kinetics for these systems.

Keywords: brown carbon; secondary organic aerosol; absorbance; aqueous aerosol; modeling

1. Introduction

Atmospheric aerosols are a major source of uncertainty in current climatological models, contributing both warming and cooling effects to the total energy balance of the planet. The contributing effects of aerosols are dependent on their composition, which contain highly variable concentrations of both inorganic and organic compounds [1,2]. Much of the organic component in aerosols is secondary organic aerosol (SOA) mass, formed indirectly through the partitioning and reaction of gas-phase volatile organic compounds (VOCs). Early literature attempting to elucidate mechanisms of SOA production emphasized the accumulation of SOA in ultrafine particles through gas-phase VOC processing into lower volatility products that uptake into the condensed phase [3,4]. However, additional relevant SOA formation mechanisms have been identified that contribute via reversible VOC uptake followed by irreversible particle-phase processing [5,6].

In aqueous aerosol systems, the composition and reactivity of water-soluble organic species dictate the prevailing aqueous-phase SOA (aqSOA) chemistry. Low-weight (C2 or C3) carbonyl-containing VOCs (CVOCs) have been identified as a relevant and ubiquitous contributor to aqSOA across a wide range of ambient atmospheric conditions [3,7,8]. Arising from the photooxidation of both biogenic and anthropogenic gas-phase emissions [9], CVOCs exhibit high partitioning into the aqueous phase due to their ability to reversibly hydrate, yielding relatively high condensed-phase concentrations compared to similarly sized or structured compounds [10]. The uptake and retention of these compounds is thought to be further enhanced by the subsequent particle-phase chemistry that occurs between other organic and electrolyte species present in the aqueous phase [11–15]. In addition, the activation and subsequent drying of cloud droplets may result in yet higher concentrations of CVOCs and derived products beyond what traditional partitioning theory for these compounds may suggest [13].

Of the CVOCs, glyoxal (G) and its reaction products has been identified to be one of the highest contributors to aqSOA mass [3,16]. Glycolaldehyde (GA) and methylglyoxal (MG), which are similarly structured to glyoxal, appear in similar environments, though in smaller quantities [16,17]. GA and MG have been proposed to undergo analogous chemistry to glyoxal in aqueous conditions [18,19]. In addition, MG has also been identified to generate comparable or greater quantities of light-absorbing dark chemistry products compared to G [20,21] and can alter the bulk-phase physical properties of aerosol systems, even in trace amounts [20,22,23]. As a result, multiple studies have explored G [6,10,16,18,19,24–29], GA [16,18,27,30,31], and MG [16,19,21,26,27,29] reaction systems, quantifying and speciating their potential contributions to aqSOA under a variety of laboratory conditions.

In dark conditions, observed higher-mass aqSOA products in G, GA, and MG mimic solutions have been attributed to Maillard-type or aldol condensation reactions between CVOCs, primary amines, and/or ammonium ions present in the aqueous phase [14,17–19,21,24,28,29]. Recent work using supercritical fluid chromatography tandem mass spectroscopy (SFC-MS/MS) has unambiguously identified heterocyclic pyrazine and imidazole derived oligomerization products in mimic solutions containing G or GA and ammonium sulfate (AS) [18]. Similar speciation studies using HPLC-MS in MG/AS aerosol mimic solutions also identified m/z peaks consistent with methylated analogues to these heterocyclic products [20,32]. These products have all been observed [20,32] or estimated [18,32] to absorb light in the UV/visible region, classifying these compounds as brown carbon (BrC) species that may contribute to overall aerosol direct forcing effects.

Whereas the existence of these reaction products has been verified, the kinetics of the parallel reactions leading to their formation remain ambiguous and poorly understood. The chromatographic methods implemented in these mass spectroscopy speciation studies necessitate elution times on the order of tens of minutes [20,33] and require additional processing to strip inorganic components from solution before measurement [23]. As a result, the ability to use dynamic m/z signals to infer kinetic information for specific products is limited. To the authors' knowledge, no rate constant data has been reported for CVOC/AS chemistry using these methods.

Other previous kinetic studies have attempted to quantify CVOC/AS dark chemistry approximate rate constants using the UV/visible absorbance at a single reference wavelength, defining regressed values as representative of the collective dark chemistry for a given reaction system [19,21,25,28]. However, this method cannot necessarily capture the nonuniform changes in UV/visible absorbance peak intensity and location that may occur over time. Laboratory studies measuring chromophoricity in CVOC/AS mimic solutions have observed different [21,23,25,28] or opposing [34] dynamic behavior at different wavelengths under the same experimental conditions. Combined with the explicitly different and parallel dark chemistry mechanisms known to be taking place within these reaction systems, a more robust method is necessary to infer kinetic information from UV/visible absorbance data.

Here, we presented a top-down spectral decomposition approach to the modeling of dynamic UV-visible absorbance properties in CVOC/AS aerosol mimics. The decomposition of a spectrum, i.e., the representation of a spectrum containing overlapping peaks as a series of simpler functions, is a widely used approach to infer additional information in a variety of spectroscopic techniques [35,36].

By representing the overall spectra of a given aged CVOC/AS system as a series of time- and wavelength-dependent empirically fitted lineshapes, a solution's absorbance spectrum can be modeled as a function of initial chromophoric precursor concentration and solution age. This technique allows for a more holistic approach to modeling absorbance behavior, informed by a spectrum's relevant geometries, as opposed to discrete sets of absorbance values at specific wavelengths.

The presented data and model parameters specifically focus on the light-absorbing properties of G, GA, and MG in AS environments, selected due to their previously explored chromophoric properties [17,18,20,21] and notable contributions to aqueous SOA mass in aqueous aerosols [16] and cloudwater droplets [7,10].

2. Materials and Methods

2.1. Sample Collection

Aqueous aerosol mimic solutions were prepared in sets of at least eight replicates in flat-bottomed, 320-μL UV-transparent 96-well plates (Corning) by combining fixed quantities of organic-containing and ammonium sulfate stock solutions. Glyoxal and methylglyoxal stock solutions were each prepared from 40 wt% reagent solution (Sigma Aldrich) and Millipore water. Glycolaldehyde stock solutions were prepared from solid dimer (Sigma Aldrich) and Millipore water. Upon initial preparation, organic stock solutions were allowed to equilibrate at room temperature for 1–3 h to allow for reversible hydration and oligomerization phenomena to take place within the solution. AS solutions were prepared from solid crystals (Sigma Aldrich) and Millipore water. The combined mimic solutions contained 3.1 M of AS and either 5 mM (for G and MG) or 5 0mM (for GA) of organic concentrations. The AS concentration was set at 3.1 M to be representative of the high inorganic concentrations typical of ambient atmospheric aerosols while remaining consistent with other work that has used these types of mimics [21,23,37,38]. CVOC concentrations were selected to remain within detectable ranges of absorbance measurement and were kept sufficiently low to reasonably assume excess ammonium concentration relative to CVOCs in solution.

Upon mixing the mimic solutions in each well using a micropipette, the 96-well plates were immediately placed into a Biotek Synergy HT Microplate Reader, where the UV-visible absorbance in each filled well was measured at hourly intervals for six hours. Between measurements, each plate was allowed to age inside the microplate reader in air under darkened, ambient (23 °C) conditions. No additional oxidants or reactants beyond AS were used. Negligible sample mass loss (<5%) was observed over the period of measurement.

Sample absorbance spectra were measured at 1-nm resolution from a range of 200–750 nm. Bulk absorbance properties, $A(\lambda)$, were then recalculated in terms of mass absorbance coefficient (MAC), to normalize for initial organic concentration, C^{o}_{CVOC} with molecular mass MW_{CVOC} and solution pathlength, l:

$$MAC(\lambda) = \frac{A(\lambda)\ln 10}{l \cdot C^{O}_{CVOC} \cdot MW_{CVOC}} \quad (1)$$

The nominal pathlength for the 320-μL mimic solution samples in each well was determined to be 7.1 mm, calibrated against a nonreactive crystal violet dye. Well plate absorbance measurements exhibited statistically insignificant dependence on well position (p = 0.4207) across the entire UV-visible region.

2.2. Modeling Approach

The models discussed in this work aimed to decompose a given MAC spectra into a sum of curves that collectively approximate the geometry of observed values, i.e., discrete peaks with nonzero width. Gaussian ($G(\lambda)$) lineshape functions, which are commonly employed in spectral decomposition techniques [36] or as a means of estimating absorbance property simulation [39], were applied such that a given spectrum was expressed as a sum of n instances of $G(\lambda)$:

$$MAC(\lambda) \approx \sum_{i}^{n} G(\lambda) = \sum_{i}^{n} M_i \cdot \exp\left(-\left(\frac{\lambda - c_i}{\sigma_i}\right)^2\right) \quad (2)$$

These functions can be represented in terms of a single magnitude-dependent (M), width-dependent (σ), and location-dependent (c) coefficient per lineshape, yielding a total of $3n$ parameters to be fitted to a single given spectrum.

The fitted Gaussian coefficients (X) may be further expanded to represent time-dependent behavior. The initial mixing and subsequent aging of the observed mimic solutions were assumed to approximate the time-domain, step change response of a first-order system, which starts at some initial value X_c before asymptotically shifting by a steady-state offset value X, dictated by an exponential decay with time constant X_τ:

$$X(t) = X_c + X\left(1 - \exp\left(-\frac{t}{X_\tau}\right)\right) \quad (3)$$

Although it is possible to model all three Gaussian coefficients as having time-dependent behavior (resulting in a total of $9n$ variables), overfitting spectra potentially increases computational load and decreases physically grounded comparability across different modeled systems. From previous literature exploring the chromophoric properties of the CVOC/AS dark chemistry system used in this work, peak spread and approximate locations were found to remain roughly constant [17,20,21,25,28]. Therefore, applying Equation (3) to the magnitude term of Equation (2), a time- and wavelength-dependent function for MAC with $5n$ fitted coefficients was developed:

$$MAC(\lambda, t) \approx \sum_{i}^{n} \left[M_{c,i} + M_i\left(1 - \exp\left(-\frac{t}{M^{I}_{\tau,i}}\right)\right)\right] \cdot \exp\left(-\left(\frac{\lambda - c_i}{\sigma_i}\right)^2\right) \quad (4)$$

Time- and wavelength-resolved MAC values, calculated through observed well plate absorbance measurements, were fitted using MATLAB (2019b) to model equations using two, three, and four Gaussian lineshape equations (i.e., Equation (4), such that n = 2, 3, and 4.) The efficacy of fitted models was quantified through a normalized root-mean-square deviation, NRSMD, defined for a testing set of N observations as:

$$NRMSD = \frac{1}{MAC_{observed,max} - MAC_{observed,min}} \sqrt{\frac{\sum_{i}^{N}(MAC_{model,i} - MAC_{observed,i})^2}{N}} \quad (5)$$

where $MAC_{model,i}$ refers to values as estimated by each model at a specified time and wavelength and $MAC_{observed}$ refers to observed measurements of a testing set under the same conditions. Estimated values were compared against the observed MAC values, averaged across replicates for each concentration and time within a 1nm range between 200–400n m, yielding a total of 201 observations per testing set. The root-mean-square (NRMSD) deviation for each set of conditions was then normalized against the range of observed MAC values in its testing set. A lower NRMSD denotes a more generally accurate model.

3. Results

3.1. *UV-Visible Absorbance of Aqueous Mimic Solutions*

Representative sets of absorbance spectra over six hours of aging, compared against fitted base case model outputs, are shown as points and shaded regions in Figure 1. MG/AS solutions were observed to be the most chromophoric, followed by G/AS and GA/AS solutions. All three solutions exhibited an asymmetrical peak in the 200–220-nm region, with a smaller peak between 260 nm and

290 nm. Generally, solution absorbance increased with solution age, with MG changing the most and GA changing the least over the measured timeframe. The absorbance locations and relative observed magnitudes in this study were shown to be consistent with previous studies that have explored comparable CVOC/AS reaction systems [17–21].

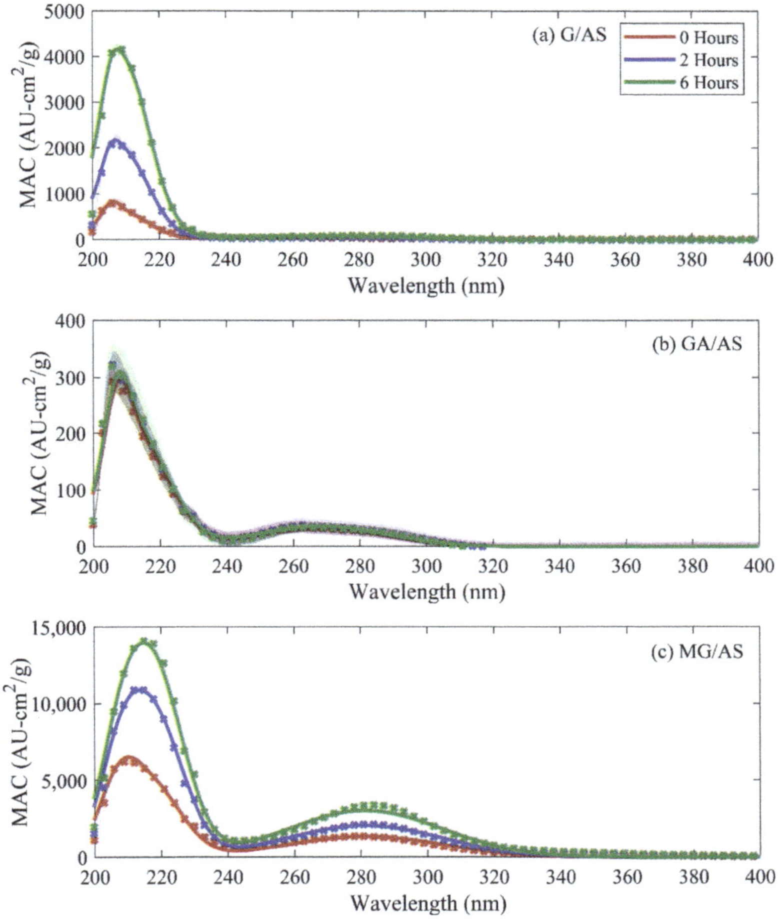

Figure 1. Observed mass absorbance coefficient (MAC) (points) and three-curve modeling output (lines) for glyoxal (**a**), glycolaldehyde (**b**), and methylglyoxal (**c**). Shaded regions denote one standard deviation ranges for observed values.

3.2. Model Comparisons

The estimated MAC for three-curve models for the three tested reaction solution systems are represented as solid lines in Figure 1. The percentage errors from the measured datasets for these outputs are shown in Figure 2. The percentage error and NRMSD information for selected simulation conditions are summarized in Table 1. Complete NRMSD data and percent error information for all combinations of CVOC/AS reaction systems and tested models can be found in Tables S1 and S2. The agreement between observed and modeled MAC was found to be high in the region between 205–300 nm, with predicted values remaining within 65% for three-curve models of all tested CVOC/AS systems and reaction times. Higher relative error was observed at the upper (>300 nm) and lower

(<205 nm) bounds of the modeled range, which was attributed to low absolute measured signal and small model-contributing sample size, respectively.

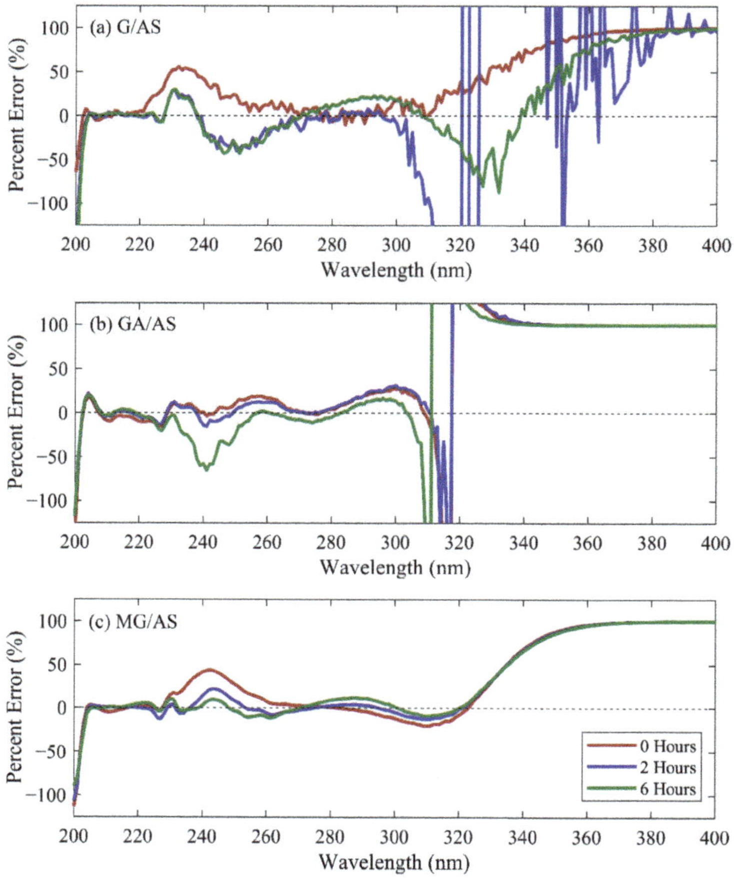

Figure 2. Percentage error vs. wavelength for the three-curve modeling scenario for glyoxal (**a**), glycolaldehyde (**b**), and methylglyoxal (**c**).

Table 1. Normalized root mean square deviation and maximum percentage error in the 205–300-nm region for selected three-curve modeling scenarios.

Species	**0 h Aging**		**2 h Aging**		**6 h Aging**	
	Max % Error	**NRMSD**	**Max % Error**	**NRMSD**	**Max % Error**	**NRMSD**
Glyoxal	55.9	0.0100	39.9	0.0117	42.6	0.0135
Glycolaldehyde	27.8	0.0125	30.5	0.0122	65.5	0.0121
Methylglyoxal	43.8	0.0146	21.6	0.0104	11.7	0.0102

Model accuracy (i.e., NRMSD) demonstrated statistically insignificant ($p = 0.9936$) dependence with respect to simulation time. NRMSD demonstrated a statistically significant ($p < 0.0001$) dependence on reaction system, with MG/AS mimics achieving the highest accuracy of the three CVOCs modeled and GA/AS achieving the lowest, trending with their relative levels of chromophoricity.

An analysis of variance (ANOVA) that compared models with different numbers of lineshapes in each CVOC/AS reaction system at each observed time resulted in statistically higher NRMSD for two-curve models, with relative error values higher in the asymmetry of the peaks in the 205–220-nm region. However, the three- and four-curve models shared a separate Tukey grouping, indicating comparable model performance between the two models. NRMSD values and multiple comparison plots from this analysis are summarized in Figure S1. As additional lineshape functions were implemented into a model equation, it was expected that incremental gains in model accuracy would diminish, while computational resources required for evaluation would increase. Therefore, it can be stated that for the CVOC/AS reaction systems examined in this work, the three-curve model provided a broadly applicable balance between fidelity to observed data and model simplicity.

4. Inference of Kinetics from Fitted Time Constants

The fitting of multiple peaks within a given absorbance spectrum suggests the potential to infer the dynamic behavior of multiple species based on the time constants of the decomposed peaks. These peaks may also be attributable to specific chromophoric compounds or functional groups using a combination of literature absorbance information for relevant families of oligomerization products and via Woodward–Fieser rules [40–42]. By separating the different inferred active groups within a batch of aqSOA products, additional insight regarding relative rates of formation among different competing formation pathways of CVOC/AS products can be achieved, providing further information to support existing means of aqSOA speciation (e.g., Fourier-transform infared spectroscopy (FTIR), nuclear magnetic resonance spectroscopy (NMR), and mass spectroscopy).

The formation of chromophoric species in CVOC/AS systems, tracked to a single wavelength, has been modeled via first order kinetics with respect to both initial organic concentration and inorganic concentration [19,21]. Of note, Sareen et al. [21] modeled the formation of chromophoric products derived from MG as following first-order kinetics through parallel ammonium- and acid-driven mechaniss. Whereas values for rate constants of both inorganic mechanisms were approximated in the previous study, it was found that the ammonium pathway rate constant exceeded the acid pathway in high-salinity aerosol mimic solutions such that only the ammonium rate constant was necessary to estimate product formation [21]. Furthermore, using an AS concentration in excess in a reaction system enables the simplification of reaction kinetics to overall pseudo-first order. Following this, and applying an analogous approach across the other CVOC/AS systems, a time constant X_τ can be connected to rate constants k^I and initial ammonium concentration, i.e.,

$$X_\tau^I = \frac{1}{\left[NH_4^+\right]k^I} \quad (6)$$

Potential product structures or functional groups, based on comparable peak locations observed in previous literature, along with inferred rate constants for these products assuming first order-kinetics (vis. Equation (6) and using an ammonium ion concentration of 6.2M), are summarized in Table 2.

Table 2. Proposed structures and first-order rate constants for selected models, inferred from fitted peak locations and time constants.

Species	Fitted Peak Location (nm)	Potential Structures or Functional Groups	Inferred Rate Constant, k^I (M^{-1} h^{-1})
G (2 curves)	208	α-carbonyls [25], pyrazine derivatives [18], imidazole derivatives [18,43,44]	0.018
	271	Pyrazine derivatives [18], aldol condensation products [28], imidazole products [43,44]	_[1]
G (3 curves)	206	α-carbonyls [25], pyrazine derivatives [18], imidazole derivatives [18,43,44]	0.038
	209	α-carbonyls [25], pyrazine derivatives, imidazole derivatives [18,43]	0.016
	267	Pyrazine derivatives [18], aldol condensation products [28], imidazole products [43,44]	_[1]
GA (3 curves)	207	α-carbonyls [25], pyrazine derivatives, imidazole derivatives [18,43]	_[1]
	213	Imidazole derivatives [18,43]	0.297
	269	Pyrazine derivatives [18]	0.092
MG (3 curves)	207	α-carbonyls [25], methylimidazole derivatives [18]	_[1]
	217	Methyl ketones [21], dimethylpyrazine derivatives [18]	0.067
	281	Dimethylpyrazine derivatives [18], aldol condensation products [21], methylimidazole derivatives [34]	0.022
MG (4 curves)	207	α-carbonyls [25], methylimidazole derivatives [18]	_[1]
	218	Methyl ketones [21], dimethylpyrazine derivatives [18]	0.066
	281	Dimethylpyrazine derivatives [18], aldol condensation products [21], methylimidazole derivatives [34]	0.016
	341	Aldol condensation products [19,21,30]	0.056

[1] Inferred rate constants do not have physical meaning and are therefore assumed to have negligible kinetics (i.e., negligible offset and/or rate of change over modeled time.).

4.1. Glyoxal + AS

Decomposed model spectra for the three-curve model of the G/AS system are shown in Figure 3. Of the two major peaks comprising the G/AS system, only the peak in the low 200-nm region exhibited dynamics outside of observational noise. This curve was represented by the three-curve composition as two overlapping lineshapes at 206 nm and 209 nm. Several chromophoric compounds have been identified in this region. Of note, imidazoles and its derivatives (e.g., imidazole carboxaldehyde) are measurable in ambient systems [45] and have been positively identified to strongly absorb in the low 200-nm range in acidic systems [43,44]. Pyrazines have also been proposed as G/AS dark chemistry products that absorb in this region, though they also contribute to absorbance in the 255–275-nm range [18]. In addition, glyoxal itself has been identified to absorb in this region, owing to its α-carbonyl functional groups [25,44]. Whereas the proximity of the two fitted lineshape functions limits the ability to reliably assign specific compounds or functional groups to each wavelength, the decomposition of the spectra clearly identifies two distinct sets of dynamics that may help distinguish between two proposed reaction pathways that contribute chromophores at these wavelengths.

The overlap in peaks may result in conflation of multiple reaction products contributing to a single value of reaction product MAC. Whereas older studies have used absorbance at a single reference wavelength to model kinetics [19,21,25,46], the resulting inferred rate constants resulted in a weighted average value that did not completely capture the parallel kinetics in play (Figure 3b). In the case of the G/AS solutions modeled in this work, using solely the MAC values at 208 nm yielded a fitted rate constant of 0.020 $M^{-1}h^{-1}$. This value is comparable to the peak at 208 nm in the two-curve model of the G/AS system (Table 2) and the 209-nm peak in the three-curve model. However, this value also slightly overstated their kinetics in trying to capture the behavior of the faster dynamics occurring in the 206-nm region, which were lost in the integration into this singular rate constant.

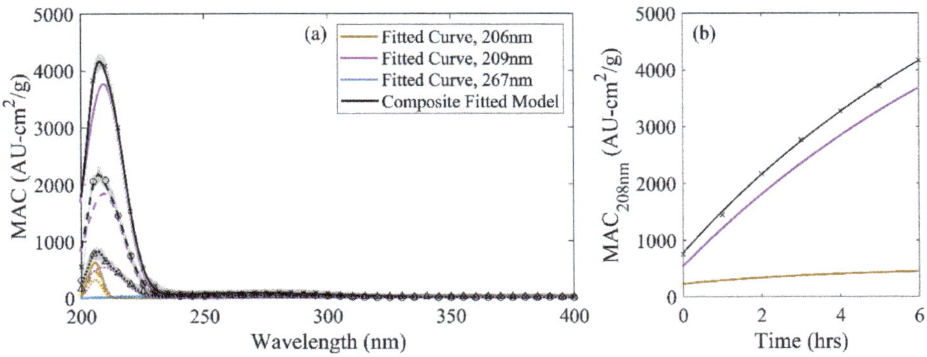

Figure 3. (**a**) Decomposed, three-curve G/AS spectrum: (a) Across wavelength at 0 h (dotted), 2 h (dashed), and 6 h (solid) of solution aging compared against observations. (points) Shaded regions denote one standard deviation ranges for observed values. (**b**) Observed MAC at 208 nm (points) compared against the overall MG/AS three-curve model (black) and contributing lineshape functions.

4.2. *Glycolaldehyde + AS*

The three-curve model spectral decomposition for the GA/AS reaction system is shown in Figure 4. This solution demonstrates the smallest absorbance dynamics of the three reaction systems tested, with MAC values increasing by less than 5% over six hours of solution aging in the entire range of wavelengths measured and modeled. Powelson et al. [17] observed similar behavior, with relatively quick but small increases in solution absorbance. Whereas changes were minor compared to those of the G/AS system, it has been demonstrated that several of the same heterocyclic, dark chemistry products that absorb in the low 200-nm range from the G/AS system are capable of forming in GA/AS solutions [18]. Although the heterocyclic species identified were ascertained from dried mimic solutions rather than aged samples, the inferred locations from aged solutions were expected to be largely consistent with these similarities to the G/AS system. The regressed locations of lineshape functions, summarized in Table 2, align with this assertion. However, different rate constants were estimated compared to the G/AS system, with negligible kinetics occurring from its 207-nm curve and nonnegligible kinetics in the 269-nm region. The similarities in contributing species with differing kinetics implies that, while comparable chromophores and contributing species are possible in the GA/AS system compared to G/AS, prevailing chemistry may vary between them.

As with G/AS, the two overlapping peaks in the low 200-nm region presented a potential misrepresentation of kinetics due to multiple contributing sources. The two contributing lineshape functions, while representative of distinctly different kinetics, had more comparable magnitudes than those in the G/AS system (Figure 4b). The local maximum in this region shifted from 207–209 nm over the observed timeframe. Using MAC values at 208 nm to infer kinetics, an effective rate constant of 0.3537 $M^{-1}h^{-1}$ was attained, which is more representative of the shifting relative contribution of the two lineshape functions rather than their actual kinetics.

The smaller signal-to-noise ratio present in the GA/AS system presents additional challenges in the inference of kinetics, as GA/AS dark chemistry products are significantly less chromophoric than G/AS or MG/AS under similar concentrations and reaction conditions. Inferred rate constants using single-wavelength, time-resolved MAC were necessarily higher in variance, yielding effective rate constants ranging from 0.076–0.430 $M^{-1}h^{-1}$ for MAC values between 205–215 nm. As a spectral decomposition method can regress values using the data from an entire spectrum, a properly constructed set of lineshape functions may be used to infer kinetic information in a more broadly informed fashion for a given set of time-resolved data.

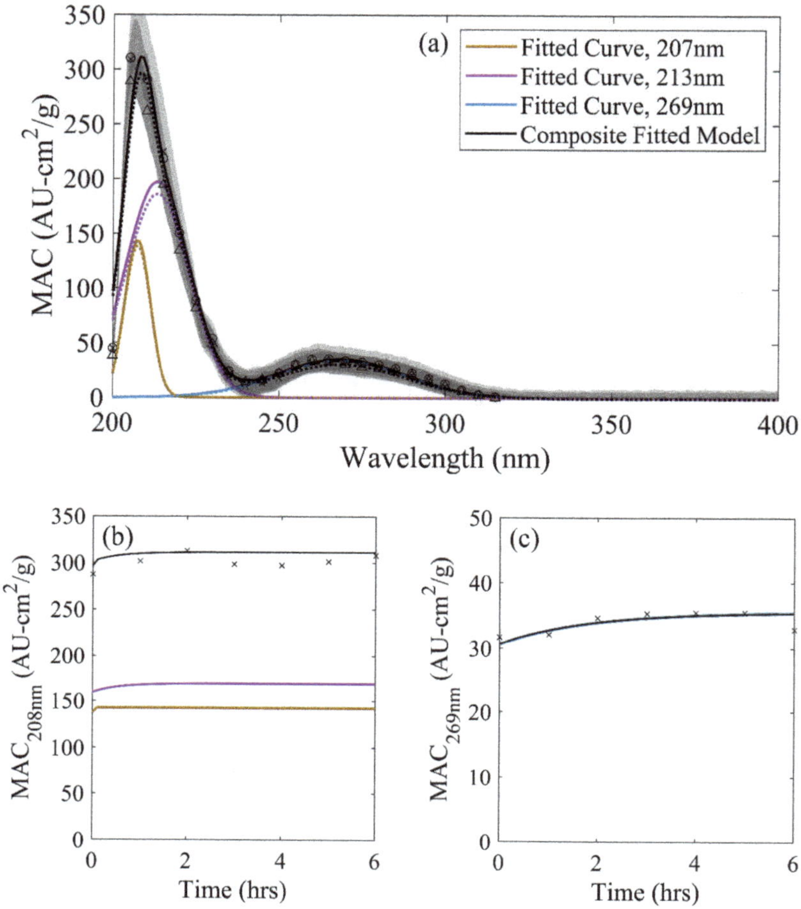

Figure 4. (**a**) Decomposed, three-curve GA/AS spectrum: (a) Across wavelength at 0 h (dotted), 2 h (dashed), and 6 h (solid) of solution aging compared against observations (points). Shaded regions denote one standard deviation ranges for observed values. (**b**) Observed MAC at 208 nm (points) compared against the overall GA/AS three-curve model (black) and contributing lin eshape functions. (**c**) Observed MAC at 269 nm (points) compared against the overall GA/AS three-curve model (black) and contributing lineshape functions.

4.3. Methylglyoxal + AS

Decomposed spectra for the MG/AS system are shown in Figure 5. The larger of the two peaks in the MG/AS reaction system exhibited a local maximum shifting from 210 nm to 215 nm over the observed period and was decomposed into two Gaussian lineshapes at 207 nm and 217 nm. Previous observations for this region are limited due to signal saturation of absorbance measurements in earlier studies [17,19,21]. However, the α-carbonyl groups, such as those from MG itself, and methylated analogues to the heterocyclic reaction products of the G/AS and GA/AS reaction systems have been identified in the MG/AS system [18,32], and are likely to absorb at these respective wavelengths [18]. The fitted 207-nm peak exhibited negligible dynamics as in GA/AS solutions (Figure 5b), suggesting that the chromophores contributing to the measurable increase in MAC in the G/AS system were formed via mechanisms unique to its reaction system. The modeled 207-nm peak was estimated to be relatively noncontributory to absorbance at 215 nm, the local maximum of this peak. Inferred rate constants using MAC values at 215 nm only yielded comparable (0.068 $M^{-1}h^{-1}$) inferred kinetics to those regressed via spectral decomposition (Table 2).

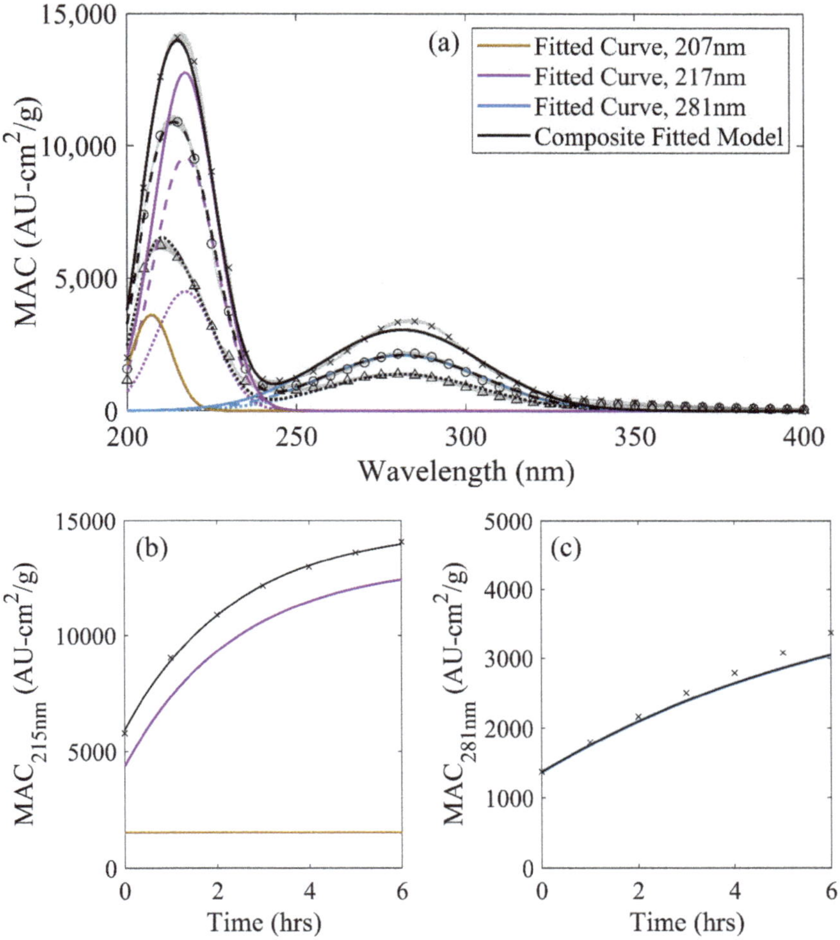

Figure 5. (**a**) Decomposed, three-curve MG/AS spectrum: (a) Across wavelength at 0 h (dotted), 2 h (dashed), and 6 h (solid) of solution aging compared against observations (points). Shaded regions denote one standard deviation ranges for observed values. (**b**) Observed MAC at 215 nm (points) compared against the overall MG/AS three-curve model (black) and contributing lineshape functions. (**c**) Observed MAC at 281 nm (points) compared against the overall MG/AS three curve model (black) and contributing lineshape functions.

The smaller peak in the 281-nm region has been observed in multiple studies examining the MG/AS system [17,19–21,34,47,48] and has been largely attributed to oligomerization products containing imidazole carbonyl variants and methylglyoxal [20,34]. Methylpyrazine products have also been identified in MG/AS solution mimics, and are consistent with HPLC-MS measurements that observed m/z values for specific chromophoric MG/AS dark chemistry products [20,32]. The single Gaussian lineshape that represents the 281-nm region exhibits fairly good agreement with observation, remaining within 11% of observed values for the three-curve model (vis. Figures 1c and 5c), though estimates slightly faster kinetics than rate constants regressed using absorbance at 281 nm

(0.0131 $M^{-1}h^{-1}$). Notably, these estimates are higher than previously reported rate constants for the MG/AS system, though literature rate constants inferred from references using absorbance information at this wavelength has yielded widely varying values, spanning multiple orders of magnitude [21,47,48]. It is likely that additional features to the MG/AS system which were not constrained in these studies, such as surfactant behavior [21,23] or the tendency to "salt out" of solution [21,27], may reduce the comparability of these values in bulk systems.

Additional literature exploring the chromophoric properties of the MG/AS system have also noted a shoulder in the 350-nm region at higher organic concentrations or duration of solution aging, which were attributable to further oligomerized aldol condensation products [19,30]. Whereas the implementation of an additional lineshape function in this region led to statistically insignificant improvement to model accuracy, a rate constant may still be inferred, as reported in Table 2. The relatively small magnitude of solution MAC at these wavelengths imply that, as observed, their overall contribution to overall solution light-absorbing properties may not be significant until sufficiently high concentrations of these products have accumulated.

Whereas the three- and four-curve Gaussian models were able to achieve good agreement with measured MAC, some studies that have measured absorbance properties of MG/AS mimic solutions aged on the order of days [20,21] have noted an asymmetry in this peak at longer reaction times. As additional features become more visible in this overall UV/visible spectrum, it is likely that the implementation of additional lineshapes into its decomposition will yield higher accuracy at long simulation times, potentially elucidating additional kinetics that occur outside the time frame utilized in this work.

5. Atmospheric Implications

The dynamic measurement of UV-visible absorbance of CVOC/AS systems yields spectra that change in relative peak heights and maximum absorbance wavelengths over their lifetimes, providing an impetus to create modeling approaches that capture these changes more accurately than via the scaling of an individual reference spectrum. The decomposition of these spectra into a series of Gaussian curves presents a flexible and powerful means of representing a range of dynamic behaviors into a relatively small number of fitted parameters. In addition, this method also presents a potential method to infer information about structure or chemical kinetics in CVOC/AS systems where multiple competing phenomena contribute to a single absorbance reading in the UV-visible region. Using this type of analysis, the entire UV/visible spectra can inform kinetic modeling rather than discrete values, enabling the identification of additional potential products and parallel reaction pathways from underlying peaks that may not be immediately apparent upon first inspection.

The spectral decomposition approach used in this study can be applied to other BrC reaction systems to better understand contributing chromophores and their respective dynamics. The number of Gaussian curves required to accurately represent light-absorbing properties is likely to differ from system to system. For products that have varied or complex spectra, such as those observed in terpene-derived aqSOA products [34], additional lineshape functions will necessarily be required to capture their dynamics. Compared to the model equations used in this work, the resultant functions will require larger amounts of training data to reliably and unambiguously infer coefficients. Data sampling rates and resolution should therefore be considered when assessing the viability of a decomposition approach to model solution kinetics.

UV/visible absorbance data is an easily collected and relatively noninvasive method of measuring bulk solution properties and is therefore a useful proxy for inferring concentration-dependent phenomena such as chemical kinetics. Whereas the derived coefficients using the approach described in this work could theoretically be used to explicitly predict the dynamic absorbance behavior of CVOC aerosol mimic reaction systems, the direct applicability of these results to the absorbance of atmospheric aerosols is limited. Without high surface-volume ratios, heterogeneous chemistry and cross-reactions of surfactant species [23] will not be accurately captured in mimic solutions, resulting in potential differences in prevailing chromophore formation. In addition, the changes to the condensed phase itself due to shifts in the temperature, pressure, and humidity surrounding an aerosol in its lifetime would result in partitioning and phase change phenomena that were not factored into these studies. To this end, when using this decomposition technique to infer information about the light absorbing properties of brown carbon in aqueous systems, it should be emphasized that the primary purpose of this modeling technique is to model rate information of a reaction system, rather than the chromophoricities of the products themselves.

Supplementary Materials: The following are available online at http://www.mdpi.com/2073-4433/11/4/358/s1, Table S1: Normalized Root Mean Square Deviation (NRMSD) values for simulations performed at varying times, numbers of Gaussian lineshapes, and CVOC/AS reaction systems, Table S2: Maximum percentage error in the 205–300nm region for simulations performed at varying times, numbers of Gaussian lineshapes, and CVOC/AS reaction systems, Figure S1: Multiple comparison test of population marginal means for the NRMSD of different CVOC/AS reaction systems and number of Gaussian curves used in model regression, across all times (0–6 h).

Author Contributions: Formal analysis, M.F., S.M. and J.L.W.; Investigation, J.L.W.; Methodology, M.F., S.M., N.F. and Z.D.; Writing—original draft, J.L.W.; Writing—review & editing, J.L.W. All authors have read and agreed to the published version of the manuscript.

Funding: This research received no external funding.

Acknowledgments: The authors thank Melissa Galloway and Aseel Bala for useful and constructive discussion in model development and chemistry relevant to the observed systems.

Conflicts of Interest: The authors declare no conflict of interest.

References

1. McNeill, V.F. Aqueous organic chemistry in the atmosphere: Sources and chemical processing of organic aerosols. *Environ. Sci. Technol.* **2015**, *49*, 1237–1244. [CrossRef] [PubMed]
2. Herrmann, H.; Schaefer, T.; Tilgner, A.; Styler, S.A.; Weller, C.; Teich, M.; Otto, T. Tropospheric Aqueous-Phase Chemistry: Kinetics, Mechanisms, and Its Coupling to a Changing Gas Phase. *Chem. Rev.* **2015**, *115*, 4259–4334. [CrossRef] [PubMed]
3. Ervens, B.; Turpin, B.J.; Weber, R.J. Secondary organic aerosol formation in cloud droplets and aqueous particles (aqSOA): A review of laboratory, field and model studies. *Atmos. Chem. Phys.* **2011**, *11*, 11069–11102. [CrossRef]
4. Faust, J.A.; Wong, J.P.S.; Lee, A.K.Y.; Abbatt, J.P.D. Role of Aerosol Liquid Water in Secondary Organic Aerosol Formation from Volatile Organic Compounds. *Environ. Sci. Technol.* **2017**, *51*, 1405–1413. [CrossRef]
5. Carlton, A.G.; Turpin, B.J. Particle partitioning potential of organic compounds is highest in the Eastern US and driven by anthropogenic water. *Atmos. Chem. Phys.* **2013**, *13*, 10203–10214. [CrossRef]
6. Volkamer, R.; San Martini, F.; Molina, L.T.; Salcedo, D.; Jimenez, J.L.; Molina, M.J. A missing sink for gas-phase glyoxal in Mexico City: Formation of secondary organic aerosol. *Geophys. Res. Lett.* **2007**, *34*, L19807. [CrossRef]
7. Ervens, B.; Carlton, A.G.; Turpin, B.J.; Altieri, K.E.; Kreidenweis, S.M.; Feingold, G. Secondary organic aerosol yields from cloud-processing of isoprene oxidation products. *Geophys. Res. Lett.* **2008**, *35*, L02816. [CrossRef]
8. Hecobian, A.; Zhang, X.; Zheng, M.; Frank, N.; Edgerton, E.S.; Weber, R.J. Water-Soluble Organic Aerosol material and the light-absorption characteristics of aqueous extracts measured over the Southeastern United States. *Atmos. Chem. Phys.* **2010**, *10*, 5965–5977. [CrossRef]
9. Fu, T.-M.; Jacob, D.J.; Wittrock, F.; Burrows, J.P.; Vrekoussis, M.; Henze, D.K. Global budgets of atmospheric glyoxal and methylglyoxal, and implications for formation of secondary organic aerosols. *J. Geophys. Res.* **2008**, *113*, D15303. [CrossRef]
10. Ervens, B.; Volkamer, R. Glyoxal processing by aerosol multiphase chemistry: Towards a kinetic modeling framework of secondary organic aerosol formation in aqueous particles. *Atmos. Chem. Phys.* **2010**, *10*, 8219–8244. [CrossRef]
11. Liggio, J.; Li, S.; McLaren, R. Reactive uptake of glyoxal by particulate matter. *J. Geophys. Res.* **2005**, *110*, D10304. [CrossRef]

12. Galloway, M.M.; Chhabra, P.S.; Chan, A.W.H.; Surratt, J.D.; Flagan, R.C.; Seinfeld, J.H.; Keutsch, F.N. Glyoxal uptake on ammonium sulphate seed aerosol: Reaction products and reversibility of uptake under dark and irradiated conditions. *Atmos. Chem. Phys.* **2009**, *9*, 3331–3345. [CrossRef]
13. De Haan, D.O.; Corrigan, A.L.; Tolbert, M.A.; Jimenez, J.L.; Wood, S.E.; Turley, J.J. Secondary Organic Aerosol Formation by Self-Reactions of Methylglyoxal and Glyoxal in Evaporating Droplets. *Environ. Sci. Technol.* **2009**, *43*, 8184–8190. [CrossRef] [PubMed]
14. De Haan, D.O.; Hawkins, L.N.; Kononenko, J.A.; Turley, J.J.; Corrigan, A.L.; Tolbert, M.A.; Jimenez, J.L. Formation of Nitrogen-Containing Oligomers by Methylglyoxal and Amines in Simulated Evaporating Cloud Droplets. *Environ. Sci. Technol.* **2011**, *45*, 984–991. [CrossRef]
15. Galloway, M.M.; Loza, C.L.; Chhabra, P.S.; Chan, A.W.H.; Yee, L.D.; Seinfeld, J.H.; Keutsch, F.N. Analysis of photochemical and dark glyoxal uptake: Implications for SOA formation. *Geophys. Res. Lett.* **2011**, *38*. [CrossRef]
16. McNeill, V.F.; Woo, J.L.; Kim, D.D.; Schwier, A.N.; Wannell, N.J.; Sumner, A.J.; Barakat, J.M. Aqueous-Phase Secondary Organic Aerosol and Organosulfate Formation in Atmospheric Aerosols: A Modeling Study. *Environ. Sci. Technol.* **2012**, *46*, 8075–8081. [CrossRef]
17. Powelson, M.H.; Espelien, B.M.; Hawkins, L.N.; Galloway, M.M.; De Haan, D.O. Brown Carbon Formation by Aqueous-Phase Carbonyl Compound Reactions with Amines and Ammonium Sulfate. *Environ. Sci. Technol.* **2014**, *48*, 985–993. [CrossRef]
18. Grace, D.N.; Sharp, J.R.; Holappa, R.E.; Lugos, E.N.; Sebold, M.B.; Griffith, D.R.; Hendrickson, H.P.; Galloway, M.M. Heterocyclic Product Formation in Aqueous Brown Carbon Systems. *ACS Earth Space Chem.* **2019**, *3*, 2472–2481. [CrossRef]
19. Schwier, A.N.; Sareen, N.; Mitroo, D.; Shapiro, E.L.; McNeill, V.F. Glyoxal-Methylglyoxal Cross-Reactions in Secondary Organic Aerosol Formation. *Environ. Sci. Technol.* **2010**, *44*, 6174–6182. [CrossRef]
20. Lin, P.; Laskin, J.; Nizkorodov, S.A.; Laskin, A. Revealing Brown Carbon Chromophores Produced in Reactions of Methylglyoxal with Ammonium Sulfate. *Environ. Sci. Technol.* **2015**, *49*, 14257–14266. [CrossRef]
21. Sareen, N.; Schwier, A.N.; Shapiro, E.L.; Mitroo, D.; McNeill, V.F. Secondary organic material formed by methylglyoxal in aqueous aerosol mimics. *Atmos. Chem. Phys.* **2010**, *10*, 997–1016. [CrossRef]
22. Sareen, N.; Schwier, A.N.; Lathem, T.L.; Nenes, A.; McNeill, V.F. Surfactants from the gas phase may promote cloud droplet formation. *Proc. Natl. Acad. Sci. USA* **2013**, *110*, 2723–2728. [CrossRef] [PubMed]
23. Beier, T.; Cotter, E.R.; Galloway, M.M.; Woo, J.L. In Situ Surface Tension Measurements of Hanging Droplet Methylglyoxal/Ammonium Sulfate Aerosol Mimics under Photooxidative Conditions. *ACS Earth Space Chem.* **2019**, *3*, 1208–1215. [CrossRef]
24. Galloway, M.M.; Powelson, M.H.; Sedehi, N.; Wood, S.E.; Millage, K.D.; Kononenko, J.A.; Rynaski, A.D.; De Haan, D.O. Secondary Organic Aerosol Formation during Evaporation of Droplets Containing Atmospheric Aldehydes, Amines, and Ammonium Sulfate. *Environ. Sci. Technol.* **2014**, *48*, 14417–14425. [CrossRef]
25. Nozière, B.; Dziedzic, P.; Córdova, A. Products and Kinetics of the Liquid-Phase Reaction of Glyoxal Catalyzed by Ammonium Ions (NH_4^+). *J. Phys. Chem. A* **2009**, *113*, 231–237. [CrossRef] [PubMed]
26. Loeffler, K.W.; Koehler, C.A.; Paul, N.M.; De Haan, D.O. Oligomer Formation in Evaporating Aqueous Glyoxal and Methyl Glyoxal Solutions. *Environ. Sci. Technol.* **2006**, *40*, 6318–6323. [CrossRef]
27. Stangl, C.M.; Johnston, M.V. Aqueous Reaction of Dicarbonyls with Ammonia as a Potential Source of Organic Nitrogen in Airborne Nanoparticles. *J. Phys. Chem. A* **2017**, *121*, 3720–3727. [CrossRef]
28. Shapiro, E.L.; Szprengiel, J.; Sareen, N.; Jen, C.N.; Giordano, M.R.; McNeill, V.F. Light-absorbing secondary organic material formed by glyoxal in aqueous aerosol mimics. *Atmos. Chem. Phys. Discuss.* **2009**, *9*, 59–80. [CrossRef]
29. Woo, J.L.; Kim, D.D.; Schwier, A.N.; Li, R.; McNeill, V.F. Aqueous aerosol SOA formation: Impact on aerosol physical properties. *Faraday Discuss.* **2013**, *165*, 357–367. [CrossRef]
30. Hawkins, L.N.; Lemire, A.N.; Galloway, M.M.; Corrigan, A.L.; Turley, J.J.; Espelien, B.M.; De Haan, D.O. Maillard Chemistry in Clouds and Aqueous Aerosol As a Source of Atmospheric Humic-Like Substances. *Environ. Sci. Technol.* **2016**, *50*, 7443–7452. [CrossRef]
31. Yi, Y.; Cao, Z.; Zhou, X.; Xue, L.; Wang, W. Formation of aqueous-phase secondary organic aerosols from glycolaldehyde and ammonium sulfate/amines: A kinetic and mechanistic study. *Atmos. Environ.* **2018**, *181*, 117–125. [CrossRef]

32. Hawkins, L.N.; Welsh, H.G.; Alexander, M.V. Evidence for pyrazine-based chromophores in cloud water mimics containing methylglyoxal and ammonium sulfate. *Atmos. Chem. Phys.* **2018**, *18*, 12413–12431. [CrossRef]
33. Grace, D.N.; Sebold, M.B.; Galloway, M.M. Separation and detection of aqueous atmospheric aerosol mimics using supercritical fluid chromatography-mass spectrometry. *Atmos. Meas. Tech* **2019**, *12*, 3841–3851. [CrossRef]
34. Aiona, P.K.; Lee, H.J.; Leslie, R.; Lin, P.; Laskin, A.; Laskin, J.; Nizkorodov, S.A. Photochemistry of Products of the Aqueous Reaction of Methylglyoxal with Ammonium Sulfate. *ACS Earth Space Chem.* **2017**, *1*, 522–532. [CrossRef]
35. de Levie, R. On deconvolving spectra. *Am. J. Phys.* **2004**, *72*, 910–915. [CrossRef]
36. Barker, B.E.; Fox, M.F. Computer resolution of overlapping electronic absorption bands. *Chem. Soc. Rev.* **1980**, *9*, 143–184. [CrossRef]
37. Tang, M.; Alexander, J.M.; Kwon, D.; Estillore, A.D.; Laskina, O.; Young, M.A.; Kleiber, P.D.; Grassian, V.H. Optical and Physicochemical Properties of Brown Carbon Aerosol: Light Scattering, FTIR Extinction Spectroscopy, and Hygroscopic Growth. *J. Phys. Chem. A* **2016**, *120*, 4155–4166. [CrossRef]
38. Li, Z.; Schwier, A.N.; Sareen, N.; McNeill, V.F. Reactive processing of formaldehyde and acetaldehyde in aqueous aerosol mimics: Surface tension depression and secondary organic products. *Atmos. Chem. Phys.* **2011**, *11*, 11617–11629. [CrossRef]
39. Frisch, M.J.; Trucks, G.W.; Schlegel, H.B.; Scuseria, G.E.; Robb, M.A.; Cheeseman, J.R.; Scalmani, G.; Barone, V.; Petersson, G.A.; Nakatsuji, H.; et al. *Gaussian 16*; Gaussian, Inc.: Wallingford, CT, USA, 2016.
40. Woodward, R.B. Structure and the Absorption Spectra of α,β-Unsaturated Ketones. *J. Am. Chem. Soc.* **1941**, *63*, 1123–1126. [CrossRef]
41. Liljefors, T.; Allinger, N.L. Conformational analysis. 128. The Woodward-Fieser rules and α,β-unsaturated ketones. *J. Am. Chem. Soc.* **1978**, *100*, 1068–1073. [CrossRef]
42. Fieser, L.F.; Fieser, M.; Rajagopalan, S. Absorption Spectroscopy And The Structures Of The Diosterols. *J. Org. Chem.* **1948**, *13*, 800–806. [CrossRef]
43. Ackendorf, J.M.; Ippolito, M.G.; Galloway, M.M. pH Dependence of the Imidazole-2-carboxaldehyde Hydration Equilibrium: Implications for Atmospheric Light Absorbance. *Environ. Sci. Technol. Lett.* **2017**, *4*, 551–555. [CrossRef]
44. Maxut, A.; Nozière, B.; Fenet, B.; Mechakra, H. Formation mechanisms and yields of small imidazoles from reactions of glyoxal with NH4(+) in water at neutral pH. *Phys. Chem. Chem. Phys.* **2015**, *17*, 20416–20424. [CrossRef]
45. Teich, M.; Van Pinxteren, D.; Kecorius, S.; Wang, Z.; Herrmann, H. First Quantification of Imidazoles in Ambient Aerosol Particles: Potential Photosensitizers, Brown Carbon Constituents, and Hazardous Components. *Environ. Sci. Technol.* **2016**, *50*, 1166–1173. [CrossRef]
46. Sareen, N.; Moussa, S.G.; McNeill, V.F. Photochemical Aging of Light-Absorbing Secondary Organic Aerosol Material. *J. Phys. Chem. A* **2013**, *117*, 2987–2996. [CrossRef]
47. Sedehi, N.; Takano, H.; Blasic, V.A.; Sullivan, K.A.; De Haan, D.O. Temperature- and pH-dependent aqueous-phase kinetics of the reactions of glyoxal and methylglyoxal with atmospheric amines and ammonium sulfate. *Atmos. Environ.* **2013**, *77*, 656–663. [CrossRef]
48. Gao, Y.; Zhang, Y. Optical properties investigation of the reactions between methylglyoxal and glycine/ammonium sulfate. *Spectrochim. Acta Part A Mol. Biomol. Spectrosc.* **2019**, *215*, 112–121. [CrossRef]

atmosphere

MDPI

Article

Impact of Aerosol-Cloud Cycling on Aqueous Secondary Organic Aerosol Formation

William G. Tsui [1], Joseph L. Woo [2] and V. Faye McNeill [1,*]

1 Department of Chemical Engineering, Columbia University, New York, NY 10027, USA; wgt2105@columbia.edu
2 Department of Chemical and Biomolecular Engineering, Lafayette College, Easton, PA 18042, USA; wooj@lafayette.edu
* Correspondence: vfm2103@columbia.edu; Tel.: +1-212-854-2869

Received: 26 September 2019; Accepted: 28 October 2019; Published: 31 October 2019

Abstract: Chemical processing of organic material in aqueous atmospheric aerosols and cloudwater is known to form secondary organic aerosols (SOA), although the extent to which each of these processes contributes to total aerosol mass is unclear. In this study, we use GAMMA 5.0, a photochemical box model with coupled gas and aqueous-phase chemistry, to consider the impact of aqueous organic reactions in both aqueous aerosols and clouds on isoprene epoxydiol (IEPOX) SOA over a range of pH for both aqueous phases, including cycling between cloud and aerosol within a single simulation. Low pH aqueous aerosol, in the absence of organic coatings or other morphology which may limit uptake of IEPOX, is found to be an efficient source of IEPOX SOA, consistent with previous work. Cloudwater at pH 4 or lower is also found to be a potentially significant source of IEPOX SOA. This phenomenon is primarily attributed to the relatively high uptake of IEPOX to clouds as a result of higher water content in clouds as compared with aerosol. For more acidic cloudwater, the aqueous organic material is comprised primarily of IEPOX SOA and lower-volatility organic acids. Both cloudwater pH and the time of day or sequence of aerosol-to-cloud or cloud-to-aerosol transitions impacted final aqueous SOA mass and composition in the simulations. The potential significance of cloud processing as a contributor to IEPOX SOA production could account for discrepancies between predicted IEPOX SOA mass from atmospheric models and measured ambient IEPOX SOA mass, or observations of IEPOX SOA in locations where mass transfer limitations are expected in aerosol particles.

Keywords: secondary organic aerosol; isoprene epoxydiol; aqueous aerosol; cloudwater

1. Introduction

Aerosols are known to affect human health, air quality, and climate and are ubiquitous in the atmosphere. They can also act as condensation nuclei to form cloud droplets, and evaporation of liquid water from the cloud droplets allows for cycling between cloud droplets and aqueous aerosols [1–4]. Cycling between these two aqueous regimes can occur on timescales of seconds to minutes, significantly shorter than the typical lifetime of aerosols, which is roughly one week [5,6]. The uptake of water by aerosols or the evaporation of cloud droplets has the potential to impact atmospheric chemistry by altering the partitioning of semivolatile species, or perturbing equilibrium reactions in the aqueous phase. Although the composition of aqueous aerosols can vary considerably, liquid water and aqueous organic material are a significant portion of atmospheric aerosol mass [7–9], and much of the aqueous-phase organic material that comprises aerosols is known to be secondary. The cycling between aqueous aerosols and cloudwater provides two different aqueous environments in which aqueous secondary organic aerosol material (aqSOA) can form [10,11]. Since both environments are aqueous, the same chemical processes occur in both environments. However, different processes are dominant due to the difference in water content, and therefore dilution of nonvolatile components, with aqueous aerosols having less water content per volume of air than cloudwater where clouds are present [12]. For example, relatively lower water content and generally lower pH are typical of aqueous aerosols, and high concentrations of organic material can limit the availability of oxidants. This results in dark (nonphotochemical) reactions, such as acid-catalyzed accretion reactions and organosulfate formation, being more dominant in aerosol water than in cloudwater [8]. Typical cloudwater pH can range between three and six, as measured at the Whiteface Mountain Field Station summit observatory in Wilmington, NY, USA [13,14], whereas aerosol particles tend to be considerably more acidic [15].

Isoprene is the most abundant nonmethane hydrocarbon gas in the atmosphere and a major source of aqSOA [16–18]. One of the most significant pathways of aqSOA formation from isoprene is through the gas-phase photooxidation of isoprene to form isoprene epoxydiols (IEPOX) [19,20]. IEPOX secondary organic aerosols (SOA), which in part consists of tetrols and organosulfates, is formed via the acid-catalyzed ring opening of the epoxide followed by nucleophilic addition of water or anions (e.g., HSO_4^-). In laboratory studies, IEPOX SOA has been found to form most efficiently at low pH as a result of the acid-catalyzed ring-opening step. The low pH and high nucleophile concentrations in aqueous aerosol provide a favorable environment for IEPOX SOA formation [21]. However, a number of scenarios which are believed to be common for aqueous aerosols, such as organic coatings, phase separation, or low-viscosity aerosols [22–31], are known to result in mass transfer limitations which inhibit the uptake of IEPOX into aqueous aerosols, limiting subsequent IEPOX SOA formation. Despite this, IEPOX SOA is observed ubiquitously in the environment [19,20,29,32]. Therefore, it is possible that cloudwater may be a greater source of IEPOX SOA than previously believed [8]. Although the lower acidity and sulfate concentrations of cloudwater would make the rate of IEPOX SOA formation in cloudwater lower than in aqueous aerosol, other factors, such as high liquid water content and lack of organic coating or viscosity-based mass transfer limitations, may promote SOA formation [33,34].

Currently, some regional and global models cannot accurately predict atmospheric aqueous organic material due to uncertainties in SOA production despite the availability of detailed chemical models representative of atmospheric processes [3,7,35]. Such discrepancies between models and observed measurements may be attributed, in part, to the lack of consideration of the cycling between cloudwater and aqueous aerosol in the models. Although several models have shown that SOA formation does occur in cloudwater, the extent to which SOA is produced as compared to formation in aqueous aerosol is unclear since very few models represent both processes [8,36–39]. In this work, we aim to compare IEPOX SOA production in both cloudwater and aqueous aerosols while considering the cycling that can occur between the two phases upon evaporation or condensation of water. We used a recently updated version of GAMMA (Gas-Aerosol Model for Mechanism Analysis), a photochemical box model developed by the McNeill group, to quantify IEPOX SOA formation. We previously used GAMMA and a related model, simpleGAMMA, to predict IEPOX SOA in aqueous aerosols, and the results were in good agreement with ambient measurements [21,40–43]. The recent update to GAMMA includes uptake of IEPOX and subsequent formation of SOA in cloudwater, allowing for the direct comparison of the efficiency of IEPOX SOA formation in both aqueous environments, and captures the coupled effect of the two modes of processing in a single simulation.

2. Methods

In this study, all simulations are based on conditions taken from ambient measurements at Whiteface Mountain and performed in GAMMA, as described in the following sections [13].

2.1. GAMMA 5.0

GAMMA is a photochemical box model developed by the McNeill group, which includes gas phase and aqueous phase species of organic material. The reactions and tracked species in GAMMA are summarized and discussed by McNeill et al. (2012) [21]. The evolution of each species is represented by:

$$\frac{dP_i}{dt} = -k_{mt,i}a_L P_i + \frac{k_{mt,i}a_L}{H_i^*}C_i + \sum_j r_{ij,gas} + E_i - D_i, \quad (1)$$

$$\frac{dC_i}{dt} = \frac{k_{mt,i}}{\mathbf{RT}}P_i - \frac{k_{mt,i}}{H_i^*\mathbf{RT}}C_i + \sum_j r_{ij,aq}, \quad (2)$$

where P_i is the concentration of species i in the gas phase, C_i is the concentration of species i in the aqueous phase, $k_{mt,i}$ is the gas-aqueous phase mass transfer coefficient, a_L is the aqueous aerosol liquid volume fraction, H_i^* is the effective Henry's Law constant, E_i is the emission rate, D_i is the deposition rate, R is the gas constant, T is temperature, and $r_{ij,gas}$ and $r_{ij,aq}$ are the gas and aqueous phase reactions of species i, respectively [44]. A list of reactions represented by $r_{ij,gas}$ and $r_{ij,aq}$ can be found in the Supporting Information provided by McNeill et al. [21] and Woo et al. [45]. As seen in Equations (1) and (2), the gas and aqueous phases of a species i are coupled in GAMMA by the mass transfer terms. In GAMMA 5.0, these equations represent the temporal evolution of aqueous species in both aerosol and cloudwater and mass transfer of gas phase species to and from aerosol and cloudwater. More detailed information regarding GAMMA, including Henry's constants used and equations calculating uptake of gas-phase species, can be found in previous work using older versions of GAMMA [21,46].

For transitions between cloudwater and aqueous aerosols, a unitless scaling factor, f_w, is calculated as the ratio of the liquid water fraction in cloudwater to the liquid water fraction in aqueous aerosol, in order to adjust aqueous phase concentrations based on the differences in the amount of liquid water available. This factor is assumed to be 5.9×10^4 for this study, determined from ambient measurements for cloudwater [13] and from E-AIM model II, a thermodynamic model which can calculate equilibrium concentrations of aqueous phase species in aerosols, for aerosol [47,48]. The liquid water fraction is assumed to be constant except at these transitions. When transitioning from aqueous aerosol to cloudwater, all condensed phase species concentrations are scaled by a factor of f_w^{-1}, except for water, which is set to 55.5 mol L^{-1}. When transitioning from cloudwater to aqueous aerosol, concentrations of condensed phase species, except water, are instead scaled by a factor of f_w. The water concentration and relative humidity are set back to the initial conditions determined from E-AIM. The aerosol particle radius is set back to the initial radius, specified as 48 nm for this work, as all simulations in this study start with aqueous aerosols before cycling. Both transitions are treated as instantaneous by GAMMA, and chemistry of evaporation, other than that due to dilution and concentration, is not considered [49,50]. Conditions that differ for aqueous aerosol and cloudwater are summarized in Table 1. All conditions shown in this table are specified for this work and can be changed based on the aqueous system of interest, except for the water concentration and relative humidity for cloudwater.

Table 1. Difference in conditions for aqueous aerosol mode and cloudwater mode for this study in GAMMA 5.0.

	Aqueous Aerosol	Cloudwater
Liquid water fraction	1.36×10^{-11} cm^3 cm^{-3}	8.0×10^{-7} cm^3 cm^{-3}
Water concentration	46.1 mol L^{-1}	55.5 mol L^{-1}
Relative humidity	80%	100%
Radius of particle or cloud droplet	48 nm	10 μm

Aqueous processing of IEPOX is represented in GAMMA as a branching reaction with 2-methyltetrols (tetrol) and IEPOX organosulfate (IEPOX OS) as products. Following Eddingsaas et al. [34], the branching ratio of 0.4 at the highest measured acidity is used for IEPOX OS for each GAMMA simulation. Additionally, 36% of the tetrols is assumed to partition to the gas phase, by taking an average of measured ambient data points from previous studies of volatility of these species [51]. IEPOX OS is assumed to be nonvolatile, and as such serves as a proxy for other potential low-volatility species such as IEPOX OS oligomers [28]. An effective Henry's Law constant of 3×10^7 mol L^{-1} atm^{-1} is used for IEPOX partitioning to the aqueous phase, based on GAMMA simulations of the laboratory data of Nguyen et al. (2014) [33]. The formation of IEPOX SOA is dependent on pH and sulfate concentration, both of which change during the aerosol-cloudwater cycling, as previously described [34,41]. Note that no mass transfer limitations are considered for the aerosol calculations, making our calculations of IEPOX SOA formation during the aerosol portion of the aerosol-cloud cycle an upper bound.

2.2. Simulation Conditions

Initial conditions were based on published conditions for Whiteface Mountain, including initial conditions for gas phase concentrations, aqueous phase concentrations, temperature, pressure, and humidity summarized in Tables S1–S3 (see Supplementary Materials) [13]. Initial conditions for aqueous phase inorganic ion concentrations were determined using E-AIM and the assumed initial aerosol pH. Each simulation was 12 h long, from 6 a.m. to 6 p.m., however, the times at which the aqueous aerosol and the cloudwater phases were present varied between runs. In the first set of simulations, aqueous aerosols were present in the first six hours and cloudwater was present in the final six hours, with the transition between the two occurring at 12 p.m. In the second set of simulations, aqueous aerosols were present in the first three hours and the last three hours, while cloudwater was present in the middle six hours. Thus, these simulations also considered cycling from cloudwater back to aqueous aerosol. The times at which GAMMA was in cloudwater mode for these two sets of simulations are shown in Figure 1. In each set of simulations, the pH was varied between subsequent runs from 1 to 4 for aqueous aerosols and 3 to 6 for cloudwater, which included typical ranges for each, but was set constant within each throughout the duration of each aerosol or cloudwater phase [8,21,38,52,53]. Photolysis rates for Whiteface Mountain were determined from Troposphere Ultraviolet and Visible (TUV) radiation model version 5.3 and were assumed to be the same throughout the cloud parcel. These rates were updated for each minute of the GAMMA simulations.

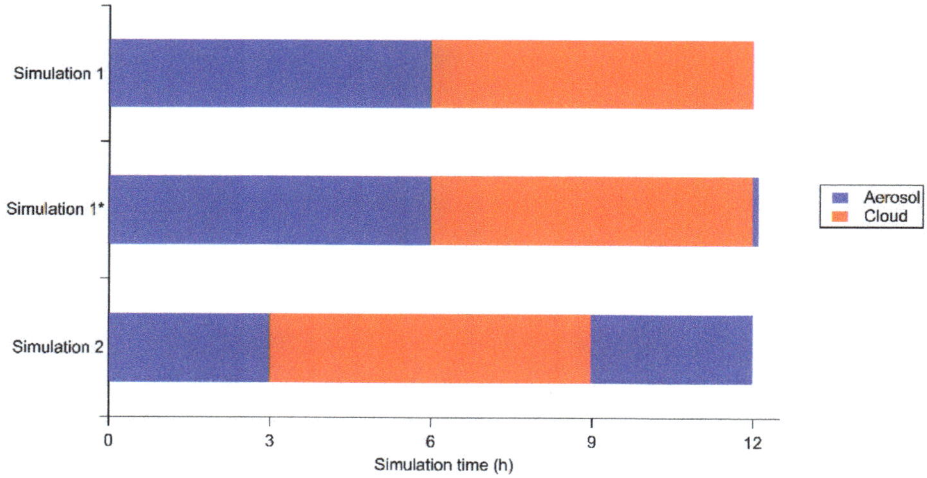

Figure 1. Times at which GAMMA is in each aqueous mode for the three sets of simulations considered in this work.

Simulation 1 ended with the condensed phase as cloudwater, whereas Simulation 2 ended with the condensed phase as aqueous aerosol. The difference in aqueous environments resulted in difficulty in comparing the final concentrations after 12 h of simulation, due to the revolatilization of some semivolatile organics that occurred upon transition from the cloud to the aerosol. For this reason, another set of simulations, called Simulation 1*, was performed where an additional minute was added onto the end of Simulation 1. The condensed phase transitioned from cloudwater to aqueous aerosol in that additional minute. The purpose of Simulation 1* was to quantify the persistent aqueous SOA created during aerosol and cloud processing in Simulation 1, such that the results of Simulations 1 and

2 could be more directly compared. GAMMA treats Henry's Law partitioning as spontaneous at every time step (Equations (1) and (2)), so one minute was sufficiently long to repartition semivolatile species to the gas phase without significant additional aqueous aerosol-phase chemical processing.

3. Results

Concentrations of aqueous phase species throughout the 12 h for all simulations are compared for different aerosol and cloud pH. Figure 2 shows the aqueous phase organic acids, carbonyl-containing VOCs (CVOCs), IEPOX (labelled IEPOXaq), and IEPOX SOA (labelled IEPOXpathway) for a Simulation 1 run with aerosol pH 1 and cloud pH 4. Organic acids include glyoxylic acid, oxalic acid, pyruvic acid, acetic acid, formic acid, glycolic acid, succinic acid, and malonic acid and their ionized forms. The CVOCs include glyoxal, methylglyoxal, glycolaldehyde, and other aldehydes.

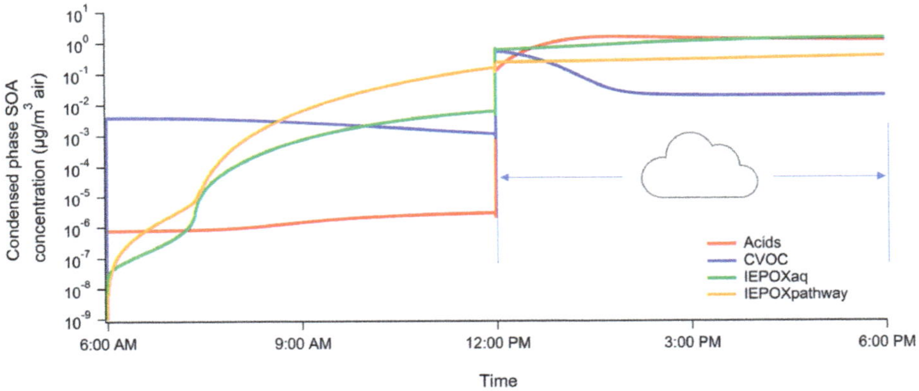

Figure 2. Organic acids, carbonyl-containing VOCs (CVOCs), isoprene epoxydiol (IEPOX), and IEPOX secondary organic aerosols (SOA) mass concentrations in Simulation 1 with aerosol pH 1 and cloud pH 4.

IEPOX and organic acids need to form in the gas phase prior to partitioning to the aerosol phase, resulting in CVOCs dominating at short simulation times due to different initial rates of formation. After six hours of photochemistry in the gas and aqueous aerosol phases, IEPOX SOA largely dominates, consistent with previous studies [21,40], however, there is a sharp increase in the mass of volatile species in the aqueous phase per volume of air following the transition to cloudwater. Because the liquid water fraction for cloudwater is approximately 10^4 times greater than that for aqueous aerosol in this work, aqueous phase concentrations are diluted by this factor, f_w, as well, creating a large concentration gradient for mass transfer into the condensed phase. This behavior is observed for both Simulations 1 and 2 at every pH considered. After one hour of cloudwater processing, organic acids are the most abundant species of these four groups, also as expected due to the greater availability of oxidants in cloudwater [8]. However, IEPOX SOA concentrations continue to increase as well, due to ongoing aqueous-phase chemistry and isoprene emissions that persist throughout the day.

On the basis of the known dependence of IEPOX SOA formation on pH, varying cloud pH in the second half of Simulation 1 is considered and the results are shown in Figure 3. In this figure, only IEPOX SOA and organic acids are compared since they are found to be the most abundant organic species besides IEPOX, as observed in Figure 2. The concentrations of these species are plotted after the first six hours of the simulation (aqueous aerosol processing), and the organic mass is found to be 99% IEPOX SOA for aerosols of pH 1. After the initial aqueous aerosol processing, the figure shows four separate cases of cloud processing for the next six hours of the simulation, at cloudwater pH ranging from 3 to 6. Although there is some IEPOX SOA formation at cloud pH 5 and 6, due to the acid catalysis requirement of the IEPOX SOA mechanism, it is more limited than the production of organic acids via the oxidation of CVOCs. However, for more acidic cloudwater, IEPOX SOA formation increases until it makes up almost 50% of aqSOA at the completion of the simulation, for cloudwater pH 3. Furthermore, for cloudwater pH 3, less than 15% of the total mass of IEPOX SOA was formed during the aerosol phase period of the simulation, suggesting the significance of cloudwater processes for IEPOX SOA formation.

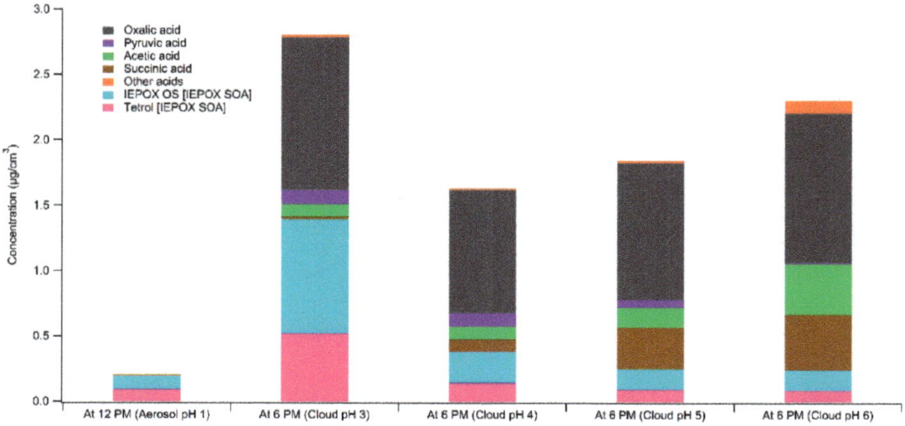

Figure 3. Organic acids and IEPOX SOA in the aqueous phase in Simulation 1 with aerosol pH 1 and cloud pH 3 to 6. The graphs for cloud pH 3 to 6 indicate composition at the end of the simulation at 6 p.m. Concentrations are given in mass per volume of air parcel.

In comparing IEPOX SOA formation at various pH for Simulation 1, the average formation rates at each pH for both aqueous aerosol and cloudwater were calculated. These results are plotted in Figure 4. As expected, due to the role of acid catalysis in aqueous-phase processing of IEPOX, increasing the pH in all cases for both cloudwater and aqueous aerosol while keeping the other constant results in a slower rate of IEPOX SOA formation [19]. Additionally, for cloud pH 5 and 6, the formation rates in cloudwater for every case are less than that of aerosol pH 1, consistent with a previous analysis [8]. However, this trend changes when cloud pH is reduced to 4, where formation of IEPOX SOA in cloudwater at all pH is comparable to formation in aerosol pH 1. At cloud pH 3, IEPOX SOA formation rates in cloudwater are approximately 10 times greater than in aerosol pH 1. Thus, due to the relatively large liquid water content of clouds, cloudwater processing can be a significant source of IEPOX SOA at pH ≤ 4, whereas stronger acidity is required for efficient IEPOX SOA formation in aqueous aerosols [21,22]. Although large IEPOX SOA production rates are observed for cloudwater, it should be noted that the rates are taken as an average over the entire six hours after cycling from aerosol. This includes the sharp increase due to rapid mass transfer of IEPOX into the aqueous phase seen shortly after the six-hour point of the simulation. Since this sharp increase does not occur again throughout the remainder of the six hours, the average formation rate is skewed high due to the beginning of the six hours. If the cloudwater phase were to run for longer than six hours, the average rate would decrease. On the other hand, aerosol-cloud cycling may occur on shorter timescales [54,55], in which case this initial dilution effect would be more prominent. Even a small amount of time of IEPOX processed in cloudwater can lead to considerable SOA growth.

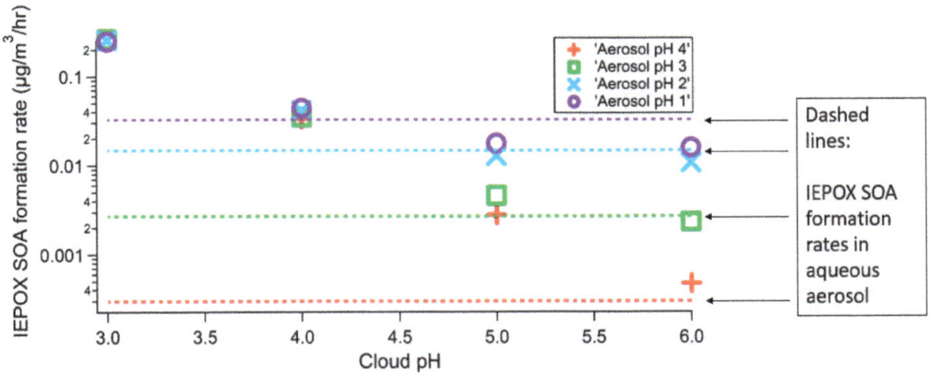

Figure 4. Comparison of IEPOX SOA formation rates in Simulation 1 for aerosol pH 1 to 4 and cloud pH 3 to 6. Rates are averaged over the entire six hours of the corresponding phase. Each individual point corresponds to the IEPOX SOA formation rate in cloudwater in the second half of the simulation with the pH of the aerosol in the first half of the simulation represented by the shape and color of the point as described in the legend. The same colors are used in the horizontal dashed lines corresponding to the formation in aqueous aerosols at the same pH in the first half of the simulation.

A similar analysis was performed for Simulation 2, where organic acids, CVOCs, IEPOX, and IEPOX SOA concentrations were compared at aerosol-cloud transitions. Figure 5, which tracks these four groups of species, shows a sharp increase in aqueous phase organic material shortly after transitioning to cloudwater, similar to Simulation 1. When cycling back to aqueous aerosol, a drop in aqueous phase organic material is observed due to the decrease in water content in the aerosol, creating a concentration gradient driving partitioning back to the gas phase. This drop is not observed for all species due to varying volatilities of aqueous-phase organics. This is more clearly seen in Figure 6, which shows the fraction of organic acids and IEPOX SOA for the following two cases of Simulation 2: (1) aqueous aerosol processing at pH 1, cloudwater at pH 4, and aqueous aerosol again at pH 1; and (2) aqueous aerosol processing at pH 1, cloudwater at pH 5, and aqueous aerosol again at pH 1. The primary difference in composition between the initial and final composition of the aqueous aerosol after cloudwater processing is the lack of more volatile acids, such as pyruvic acid, in the final composition. The species that remain in the aqueous phase have higher Henry's Law constants [21].

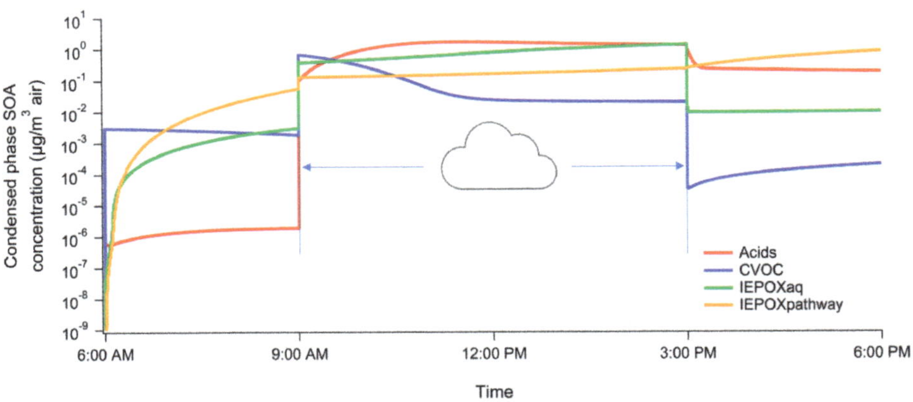

Figure 5. Organic acids, CVOCs, IEPOX, and IEPOX SOA mass concentrations in Simulation 2 with aerosol pH 1 and cloud pH 4.

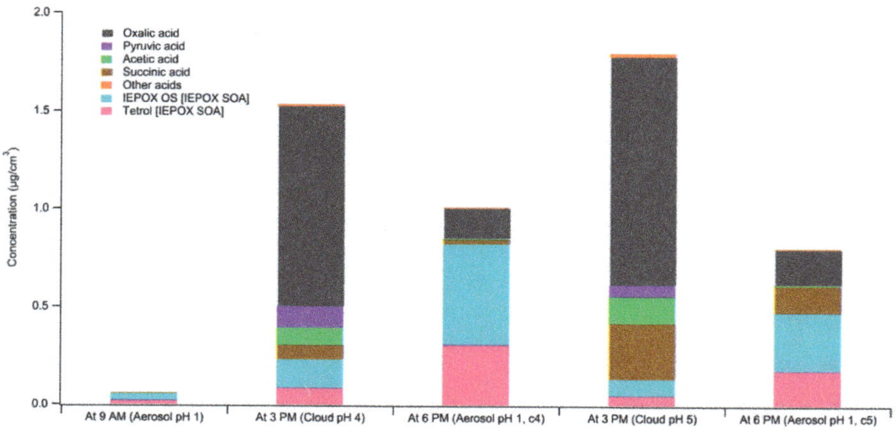

Figure 6. Organic acids and IEPOX SOA in the aqueous phase for Simulation 2 with aerosol pH 1 and cloud pH 4 or 5. The two bars labeled c4 and c5 at 6 PM correspond to Simulation 2 with cloud pH 4 and 5, respectively, as the starting and ending aerosol pH are the same while the intermediate cloud pH differs. Concentrations are given in mass per volume of air parcel.

For Simulation 2, cloud pH is observed to be the most important variable for IEPOX SOA formation. After three hours of aqueous aerosol processing, IEPOX SOA is more abundant than organic acids, as seen in Figure 6. However, the total aqSOA mass is 5.50×10^{-2} μg m^{-3}, which is insignificant relative to a mass of 1.54 μg m^{-3} and 1.80 μg m^{-3} after cloudwater processing for cloud pH 4 and 5, respectively. This is due to the comparatively low liquid water content of the aerosol phase. In this example, pH 4 or 5 for cloudwater is sufficient for IEPOX SOA formation, although organic acid formation is still more dominant.

The final mass fractions of IEPOX SOA and organic acids for Simulations 1, 1 *, and 2 are shown in Figure 7. As previously noted, the end of Simulation 1 has most of the acids produced in cloudwater still in the aqueous phase, largely due to the relatively high-water content in cloudwater, however, extending the simulation by one minute to cycle back to aqueous aerosol is sufficient for most of the volatile organic acids to partition to the gas phase, leaving primarily oxalic acid, succinic acid, and IEPOX SOA in the aqueous phase. Comparing the final mass fractions in Simulations 1* and 2, IEPOX SOA formation is greater in Simulation 2. The masses of lower-volatility organic acids are similar, but greater for Simulations 1 and 1*. Since each simulation includes the same total duration of aerosol phase and cloud phase chemistry, the difference shown in Figure 7 may be attributed to the sequence of aerosol-cloud cycling. Although isoprene emissions are assumed to be constant throughout the duration of this daytime-only simulation, the different times specified for aqueous aerosol and cloudwater results in the extent of partitioning to differ throughout the day. As more isoprene has been emitted cumulatively towards later times in the day, this may result in a greater accumulation of isoprene oxidation products when the liquid water fraction of the aqueous phase is lower later in the day, ultimately resulting in variations in IEPOX SOA formation. Another factor attributing to the difference in Simulations 1* and 2 is the variation of photolysis rates throughout the day. Photolysis rates are highest at solar noon. This is optimal for organic acid formation and gas-phase IEPOX formation since the availability of oxidants drives both processes. As seen in Figures 3, 6 and 7, cloud pH and the time of day and sequence all have a significant effect on aqueous SOA production and composition.

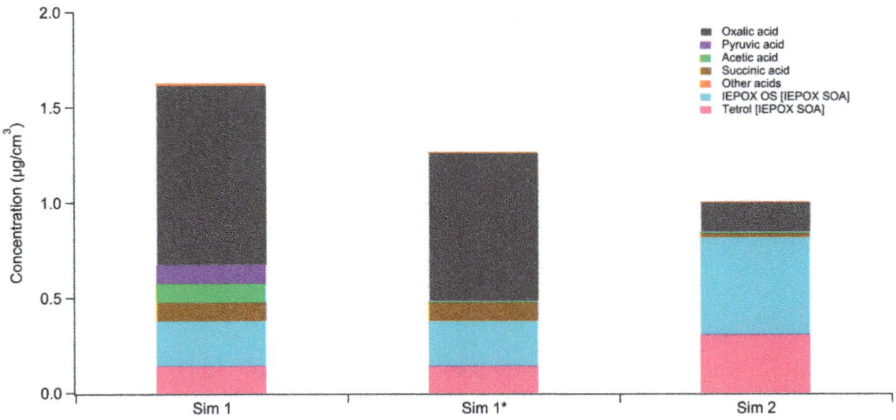

Figure 7. Mass of aqueous phase organic acids and IEPOX SOA at the end of Simulations 1, 1*, and 2 for aerosol pH 1 and cloud pH 4. Concentrations are given in terms of volume of air.

4. Atmospheric Implications

The results of this work indicate that aqueous-phase chemistry in clouds may be a more significant source of IEPOX SOA than previously believed. Production in cloudwater could account for the discrepancy between the expected efficiency of uptake of IEPOX to aqueous aerosols, given mass transfer considerations, and the ubiquity of ambient IEPOX SOA. Since no aerosol-phase mass transfer limitations were considered in our simulations, our calculations of IEPOX SOA formation in aerosol water can be considered to be an upper bound.

Due to a lack of observations of IEPOX SOA tracers in cloudwater and the high variability of liquid water content and pH in both the aerosol and cloud phases, the percentage of IEPOX SOA produced in cloudwater and aqueous aerosol remains unclear [11,38,56]. The observed lack of correlation between SOA mass and several aerosol properties, including liquid water content and pH, for ambient data may also be clarified by considering production in clouds [57]. Due to its potentially large contribution to total SOA mass, cloudwater and its efficiency in producing IEPOX SOA and other SOA should be considered when quantifying sources of organic aerosol.

Supplementary Materials: The following are available online at http://www.mdpi.com/2073-4433/10/11/666/s1, Table S1: Cloudwater simulation conditions, Table S2: Gas phase initial number concentrations specified in GAMMA, Table S3: Initial concentrations of ammonium (bi)sulfate in GAMMA, following E-AIM.

Author Contributions: V.F.M., J.L.W. and W.G.T. contributed to the GAMMA code; W.G.T. conducted the simulations discussed in this study; W.G.T. was the main author of the manuscript; V.F.M. and J.L.W. contributed in the writing process; and V.F.M. acquired the funding for this work and supervised the research.

Funding: This research was funded by the NSF-USA, grant number AGS−1546136.

Conflicts of Interest: The authors declare no conflict of interest.

References

1. Kanakidou, M.; Seinfeld, J.H.; Pandis, S.N.; Barnes, I.; Dentener, F.J.; Facchini, M.C.; Van Dingenen, R.; Ervens, B.; Nenes, A.; Nielsen, C.J.; et al. Organic aerosol and global climate modelling: A review. *Atmos. Chem. Phys.* **2005**, *5*, 1053–1123. [CrossRef]
2. Nguyen, T.K.V.; Zhang, Q.; Jimenez, J.L.; Pike, M.; Carlton, A.G. Liquid Water: Ubiquitous Contributor to Aerosol Mass. *Environ. Sci. Technol. Lett.* **2016**, *3*, 257–263. [CrossRef]
3. Jimenez, J.L.; Canagaratna, M.R.; Donahue, N.M.; Prévôt, A.S.H.; Zhang, Q.; Kroll, J.H.; DeCarlo, P.F.; Allan, J.D.; Coe, H.; Ng, N.L.; et al. Evolution of Organic Aerosols in the Atmosphere. *Science* **2009**, *326*, 1525–1529. [CrossRef] [PubMed]

4. de Gouw, J.A.; Jimenez, J.L. Organic Aerosols in the Earth's Atmosphere. *Environ. Sci. Technol.* **2009**, *43*, 7614–7618. [CrossRef] [PubMed]
5. Chuang, P.Y. Measurement of the timescale of hygroscopic growth for atmospheric aerosols. *J. Geophys. Res. Atmos.* **2003**, *108*, 4282. [CrossRef]
6. Jacob, D.J. *Introduction to Atmospheric Chemistry*; Princeton University Press: Princeton, NJ, USA, 1999; ISBN 9781680159073.
7. Hallquist, M.; Wenger, J.C.; Baltensperger, U.; Rudich, Y.; Simpson, D.; Claeys, M.; Dommen, J.; Donahue, N.M.; George, C.; Goldstein, A.H.; et al. The formation, properties and impact of secondary organic aerosol: Current and emerging issues. *Atmos. Chem. Phys.* **2009**, *9*, 5155–5236. [CrossRef]
8. McNeill, V.F. Aqueous Organic Chemistry in the Atmosphere: Sources and Chemical Processing of Organic Aerosols. *Environ. Sci. Technol.* **2015**, *49*, 1237–1244. [CrossRef]
9. Zhang, Q.; Jimenez, J.L.; Canagaratna, M.R.; Allan, J.D.; Coe, H.; Ulbrich, I.M.; Alfarra, M.R.; Takami, A.; Middlebrook, A.M.; Sun, Y.L.; et al. Ubiquity and dominance of oxygenated species in organic aerosols in anthropogenically-influenced Northern Hemisphere midlatitudes. *Geophys. Res. Lett.* **2007**, *34*, L13801. [CrossRef]
10. Herckes, P.; Valsaraj, K.T.; Collett, J.L. A review of observations of organic matter in fogs and clouds: Origin, processing and fate. *Atmos. Res.* **2013**, *132–133*, 434–449. [CrossRef]
11. Ervens, B.; Sorooshian, A.; Aldhaif, A.M.; Shingler, T.; Crosbie, E.; Ziemba, L.; Campuzano-Jost, P.; Jimenez, J.L.; Wisthaler, A. Is there an aerosol signature of chemical cloud processing? *Atmos. Chem. Phys.* **2018**, *18*, 16099–16119. [CrossRef]
12. Seinfeld, J.H.; Pandis, S.N. *Atmospheric Chemistry and Physics*; European Geophysical Society: Munich, Germany, 1998; Volume 6.
13. Lance, S.; Barth, M.; Carlton, A.; Lance, S.; Barth, M.; Carlton, A. Multiphase Chemistry: Experimental Design for Coordinated Measurement and Modeling Studies of Cloud Processing at a Mountaintop. *Bull. Am. Meteorol. Soc.* **2017**, *98*, ES163–ES167. [CrossRef]
14. Adirondack Lakes Survey Corporation-Whiteface Mountain Cloud Monitoring. Available online: http://www.adirondacklakessurvey.org/wfc.shtml (accessed on 25 September 2019).
15. Pye, H.O.T.; Nenes, A.; Alexander, B.; Ault, A.P.; Barth, M.; Clegg, S.; Collett, J.L., Jr.; Fahey, K.M.; Hennigan, C.J.; Herrmann, H.; et al. The Acidity of Atmospheric Particles and Clouds. *Atmos. Chem. Phys. Discuss.* **2019**. in review.
16. Henze, D.K.; Seinfeld, J.H. Global secondary organic aerosol from isoprene oxidation. *Geophys. Res. Lett.* **2006**, *33*, L09812. [CrossRef]
17. Guenther, A.; Karl, T.; Harley, P.; Wiedinmyer, C.; Palmer, P.I.; Geron, C. Estimates of global terrestrial isoprene emissions using MEGAN (Model of Emissions of Gases and Aerosols from Nature). *Atmos. Chem. Phys.* **2006**, *6*, 3181–3210. [CrossRef]
18. Carlton, A.G.; Wiedinmyer, C.; Kroll, J.H. A Review of Secondary Organic Aerosol (SOA) Formation from Isoprene. *Atmos. Chem. Phys.* **2009**, *9*, 4987–5005. [CrossRef]
19. Surratt, J.D.; Lewandowski, M.; Offenberg, J.H.; Jaoui, M.; Kleindienst, T.E.; Edney, E.O.; Seinfeld, J.H. Effect of acidity on secondary organic aerosol formation from isoprene. *Environ. Sci. Technol.* **2007**, *41*, 5363–5369. [CrossRef]
20. Surratt, J.D.; Chan, A.W.H.; Eddingsaas, N.C.; Chan, M.; Loza, C.L.; Kwan, A.J.; Hersey, S.P.; Flagan, R.C.; Wennberg, P.O.; Seinfeld, J.H. Reactive intermediates revealed in secondary organic aerosol formation from isoprene. *Proc. Natl. Acad. Sci. USA* **2010**, *107*, 6640. [CrossRef]
21. McNeill, V.F.; Woo, J.L.; Kim, D.D.; Schwier, A.N.; Wannell, N.J.; Sumner, A.J.; Barakat, J.M. Aqueous-Phase Secondary Organic Aerosol and Organosulfate Formation in Atmospheric Aerosols: A Modeling Study. *Environ. Sci. Technol.* **2012**, *46*, 8075–8081. [CrossRef]
22. Gaston, C.J.; Riedel, T.P.; Zhang, Z.; Gold, A.; Surratt, J.D.; Thornton, J.A. Reactive Uptake of an Isoprene-Derived Epoxydiol to Submicron Aerosol Particles. *Environ. Sci. Technol.* **2014**, *48*, 11178–11186. [CrossRef]
23. Pye, H.O.T.; Murphy, B.N.; Xu, L.; Ng, N.L.; Carlton, A.G.; Guo, H.; Weber, R.J.; Vasilakos, P.; Appel, K.W.; Budisulistiorini, S.H.; et al. On the implications of aerosol liquid water and phase separation for organic aerosol mass. *Atmos. Chem. Phys.* **2017**, *17*, 343–369. [CrossRef]

24. Schmedding, R.; Ma, M.; Zhang, Y.; Farrell, S.; Pye, H.O.T.; Chen, Y.; Wang, C.T.; Rasool, Q.Z.; Budisulistiorini, S.H.; Ault, A.P.; et al. α-Pinene-Derived organic coatings on acidic sulfate aerosol impacts secondary organic aerosol formation from isoprene in a box model. *Atmos. Environ.* **2019**, *213*, 456–462. [CrossRef]
25. Shiraiwa, M.; Berkemeier, T.; Schilling-Fahnestock, K.A.; Seinfeld, J.H.; Pöschl, U. Molecular corridors and kinetic regimes in the multiphase chemical evolution of secondary organic aerosol. *Atmos. Chem. Phys.* **2014**, *14*, 8323–8341. [CrossRef]
26. Song, M.; Liu, P.F.; Hanna, S.J.; Li, Y.J.; Martin, S.T.; Bertram, A.K. Relative humidity-dependent viscosities of isoprene-derived secondary organic material and atmospheric implications for isoprene-dominant forests. *Atmos. Chem. Phys.* **2015**, *15*, 5145–5159. [CrossRef]
27. Reid, J.P.; Bertram, A.K.; Topping, D.O.; Laskin, A.; Martin, S.T.; Petters, M.D.; Pope, F.D.; Rovelli, G. The viscosity of atmospherically relevant organic particles. *Nat. Commun.* **2018**, *9*, 956. [CrossRef]
28. D'ambro, E.L.; Schobesberger, S.; Gaston, C.J.; Lopez-Hilfiker, F.D.; Lee, B.H.; Liu, J.; Zelenyuk, A.; Bell, D.; Cappa, C.D.; Helgestad, T.; et al. Chamber-based insights into the factors controlling epoxydiol (IEPOX) secondary organic aerosol (SOA) yield, composition, and volatility. *Atmos. Chem. Phys.* **2019**, *19*, 11253–11265.
29. Hu, W.; Palm, B.B.; Day, D.A.; Campuzano-Jost, P.; Krechmer, J.E.; Peng, Z.; de Sá, S.S.; Martin, S.T.; Alexander, M.L.; Baumann, K.; et al. Volatility and lifetime against OH heterogeneous reaction of ambient isoprene-epoxydiols-derived secondary organic aerosol (IEPOX-SOA). *Atmos. Chem. Phys.* **2016**, *16*, 11563–11580. [CrossRef]
30. Zhang, Y.; Chen, Y.; Lambe, A.T.; Olson, N.E.; Lei, Z.; Craig, R.L.; Zhang, Z.; Gold, A.; Onasch, T.B.; Jayne, J.T.; et al. Effect of the Aerosol-Phase State on Secondary Organic Aerosol Formation from the Reactive Uptake of Isoprene-Derived Epoxydiols (IEPOX). *Environ. Sci. Technol. Lett.* **2018**, *5*, 167–174. [CrossRef]
31. Riva, M.; Chen, Y.; Zhang, Y.; Lei, Z.; Olson, N.E.; Boyer, H.C.; Narayan, S.; Yee, L.D.; Green, H.S.; Cui, T.; et al. Increasing Isoprene Epoxydiol-to-Inorganic Sulfate Aerosol Ratio Results in Extensive Conversion of Inorganic Sulfate to Organosulfur Forms: Implications for Aerosol Physicochemical Properties. *Environ. Sci. Technol.* **2019**, *53*, 8682–8694. [CrossRef]
32. Hu, W.W.; Campuzano-Jost, P.; Palm, B.B.; Day, D.A.; Ortega, A.M.; Hayes, P.L.; Krechmer, J.E.; Chen, Q.; Kuwata, M.; Liu, Y.J.; et al. Characterization of a real-time tracer for isoprene epoxydiols-derived secondary organic aerosol (IEPOX-SOA) from aerosol mass spectrometer measurements. *Atmos. Chem. Phys.* **2015**, *15*, 11807–11833. [CrossRef]
33. Nguyen, T.B.; Coggon, M.M.; Bates, K.H.; Zhang, X.; Schwantes, R.H.; Schilling, K.A.; Loza, C.L.; Flagan, R.C.; Wennberg, P.O.; Seinfeld, J.H. Organic aerosol formation from the reactive uptake of isoprene epoxydiols (IEPOX) onto non-acidified inorganic seeds. *Atmos. Chem. Phys.* **2014**, *14*, 3497–3510. [CrossRef]
34. Eddingsaas, N.C.; VanderVelde, D.G.; Wennberg, P.O. Kinetics and Products of the Acid-Catalyzed Ring-Opening of Atmospherically Relevant Butyl Epoxy Alcohols. *J. Phys. Chem. A* **2010**, *114*, 8106–8113. [CrossRef]
35. Heald, C.L.; Jacob, D.J.; Park, R.J.; Russell, L.M.; Huebert, B.J.; Seinfeld, J.H.; Liao, H.; Weber, R.J. A large organic aerosol source in the free troposphere missing from current models. *Geophys. Res. Lett.* **2005**, *32*, L18809. [CrossRef]
36. Blando, J.D.; Turpin, B.J. Secondary organic aerosol formation in cloud and fog droplets: A literature evaluation of plausibility. *Atmos. Environ.* **2000**, *34*, 1623–1632. [CrossRef]
37. Carlton, A.G.; Turpin, B.J.; Altieri, K.E.; Seitzinger, S.P.; Mathur, R.; Roselle, S.J.; Weber, R.J. CMAQ Model Performance Enhanced When In-Cloud Secondary Organic Aerosol is Included: Comparisons of Organic Carbon Predictions with Measurements. *Environ. Sci. Technol.* **2008**, *42*, 8798–8802. [CrossRef]
38. Ervens, B.; Turpin, B.J.; Weber, R.J. Secondary organic aerosol formation in cloud droplets and aqueous particles (aqSOA): A review of laboratory, field and model studies. *Atmos. Chem. Phys.* **2011**, *11*, 11069–11102. [CrossRef]
39. Ervens, B. Modeling the Processing of Aerosol and Trace Gases in Clouds and Fogs. *Chem. Rev.* **2015**, *115*, 4157–4198. [CrossRef]
40. Woo, J.L.; McNeill, V.F. simpleGAMMA v1.0–A reduced model of secondary organic aerosol formation in the aqueous aerosol phase (aaSOA). *Geosci. Model. Dev.* **2015**, *8*, 1821–1829. [CrossRef]
41. Budisulistiorini, S.H.; Nenes, A.; Carlton, A.G.; Surratt, J.D.; McNeill, V.F.; Pye, H.O.T. Simulating Aqueous-Phase Isoprene-Epoxydiol (IEPOX) Secondary Organic Aerosol Production During the 2013 Southern Oxidant and Aerosol Study (SOAS). *Environ. Sci. Technol.* **2017**, *51*, 5026–5034. [CrossRef]

42. Budisulistiorini, S.H.; Li, X.; Bairai, S.T.; Renfro, J.; Liu, Y.; Liu, Y.J.; Mckinney, K.A.; Martin, S.T.; Mcneill, V.F.; Pye, H.O.T.; et al. Examining the effects of anthropogenic emissions on isoprene-derived secondary organic aerosol formation during the 2013 Southern Oxidant and Aerosol Study (SOAS) at the Look Rock, Tennessee ground site. *Atmos. Chem. Phys.* **2015**, *15*, 8871–8888. [CrossRef]
43. Shrivastava, M.; Andreae, M.O.; Artaxo, P.; Barbosa, H.M.J.; Berg, L.K.; Brito, J.; Ching, J.; Easter, R.C.; Fan, J.; Fast, J.D.; et al. Urban pollution greatly enhances formation of natural aerosols over the Amazon rainforest. *Nat. Commun.* **2019**, *10*, 1–12. [CrossRef]
44. Schwartz, S.E. *Mass-Transport Considerations Pertinent to Aqueous Phase Reactions of Gases in Liquid-Water Clouds*; Jaeschke, W., Ed.; NATO ASI Series; Springer: Berlin/Heidelberg, Germany, 1986; Volume G6, pp. 425–471.
45. Woo, J.L.; Kim, D.D.; Schwier, A.N.; Li, R.; Faye McNeill, V. Aqueous aerosol SOA formation: Impact on aerosol physical properties. *Faraday Discuss.* **2013**, *165*, 357–367. [CrossRef]
46. Tsui, W.G.; Rao, Y.; Dai, H.-L.; McNeill, V.F. Modeling Photosensitized Secondary Organic Aerosol Formation in Laboratory and Ambient Aerosols. *Environ. Sci. Technol.* **2017**, *51*, 7496–7501. [CrossRef]
47. Clegg, S.L.; Brimblecombe, P.; Wexler, A.S. Thermodynamic Model of the System H + -NH_4 + -SO_{42}−NO_3 −H_2O at Tropospheric Temperatures. *J. Phys. Chem. A* **1998**, *102*, 2137–2154. [CrossRef]
48. Clegg, S.L.; Seinfeld, J.H.; Brimblecombe, P. Thermodynamic modelling of aqueous aerosols containing electrolytes and dissolved organic compounds. *J. Aerosol Sci.* **2001**, *32*, 713–738. [CrossRef]
49. Nguyen, T.B.; Lee, P.B.; Updyke, K.M.; Bones, D.L.; Laskin, J.; Laskin, A.; Nizkorodov, S.A. Formation of nitrogen- and sulfur-containing light-absorbing compounds accelerated by evaporation of water from secondary organic aerosols. *J. Geophys. Res. Atmos.* **2012**, *117*, D01207. [CrossRef]
50. De Haan, D.O.; Corrigan, A.L.; Tolbert, M.A.; Jimenez, J.L.; Wood, S.E.; Turley, J.J. Secondary organic aerosol formation by self-reactions of methylglyoxal and glyoxal in evaporating droplets. *Environ. Sci. Technol.* **2009**, *43*, 8184–8190. [CrossRef]
51. Isaacman-VanWertz, G.; Yee, L.D.; Kreisberg, N.M.; Wernis, R.; Moss, J.A.; Hering, S.V.; de Sá, S.S.; Martin, S.T.; Alexander, M.L.; Palm, B.B.; et al. Ambient Gas-Particle Partitioning of Tracers for Biogenic Oxidation. *Environ. Sci. Technol.* **2016**, *50*, 9952–9962. [CrossRef]
52. Anastasio, C.; Faust, B.C.; Rao, C.J. Aromatic carbonyl compounds as aqueous-phase photochemical sources of hydrogen peroxide in acidic sulfate aerosols, fogs, and clouds. 1. Non- phenolic methoxybenzaldehydes and methoxyacetophenones with reductants (phenols). *Environ. Sci. Technol.* **1997**, *31*, 218–232. [CrossRef]
53. Weber, R.J.; Guo, H.; Russell, A.G.; Nenes, A. High aerosol acidity despite declining atmospheric sulfate concentrations over the past 15 years. *Nat. Geosci.* **2016**, *9*, 282–285. [CrossRef]
54. Xue, H.; Feingold, G.; Xue, H.; Feingold, G. Large-Eddy Simulations of Trade Wind Cumuli: Investigation of Aerosol Indirect Effects. *J. Atmos. Sci.* **2006**, *63*, 1605–1622. [CrossRef]
55. Feingold, G.; McComiskey, A.; Rosenfeld, D.; Sorooshian, A. On the relationship between cloud contact time and precipitation susceptibility to aerosol. *J. Geophys. Res. Atmos.* **2013**, *118*, 10544. [CrossRef]
56. Sareen, N.; Waxman, E.M.; Turpin, B.J.; Volkamer, R.; Carlton, A.G. Potential of aerosol liquid water to facilitate organic aerosol formation: Assessing knowledge gaps about precursors and partitioning. *Environ. Sci. Technol.* **2017**, *51*, 3327–3335. [CrossRef]
57. Xu, L.; Guo, H.; Boyd, C.M.; Klein, M.; Bougiatioti, A.; Cerully, K.M.; Hite, J.R.; Isaacman-VanWertz, G.; Kreisberg, N.M.; Knote, C.; et al. Effects of anthropogenic emissions on aerosol formation from isoprene and monoterpenes in the southeastern United States. *Proc. Natl. Acad. Sci. USA* **2015**, *112*, 37–42. [CrossRef]

MDPI
St. Alban-Anlage 66
4052 Basel
Switzerland
Tel. +41 61 683 77 34
Fax +41 61 302 89 18
www.mdpi.com

Atmosphere Editorial Office
E-mail: atmosphere@mdpi.com
www.mdpi.com/journal/atmosphere

www.ingramcontent.com/pod-product-compliance
Lightning Source LLC
Chambersburg PA
CBHW061626210926
51850CB00009B/2445
* 9 7 8 3 0 3 6 5 7 3 2 3 6 *